ROYAL AIR FORCE

THE OFFICIAL STORY

THIS IS A WELBECK BOOK

First published in 2018 by André Deutsch Ltd.

This edition published in 2020 by Welbeck
an imprint of Welbeck Non-Fiction Ltd,
part of the Welbeck Publishing Group
20 Mortimer Street
London W1T 3JW

10 9 8 7 6 5 4 3 2 1

A CIP catalogue record for this book is available from the British
Library.

ISBN: 978 1 787 39423 0

Printed in Dubai

ROYAL AIR FORCE

THE OFFICIAL STORY

JAMES HOLLAND

WELBECK

CONTENTS

Opposite / A lone Royal Flying Corps
FE2 fighter flies over France.

1

THE
GENESIS
OF
THE
RAF

The Royal Air Force came into being on 1 April 1918, and as such was the world's first air force to operate independently of army or navy control. Its establishment came a mere fourteen years after the first-ever powered flight at Kitty Hawk in North Carolina in the United States, but, even so, Britain's politicians and military leaders had been comparatively slow to embrace aviation as an important and exciting new instrument of war.

While the Wright brothers continued their pioneering work in the United States, it was the French who were leading the way in Europe in terms of aviation, though for much of the time in ignorance of what the Wrights were doing in comparative secrecy across the Atlantic. The first French flight, spluttering and barely 60 metres (65 yards), took place in the full glare of the French public in the Bois du Boulogne in Paris in October 1906. The Germans, meanwhile, were putting much of their energy into building ever-larger dirigibles – that is, large hydrogen-filled airships. It was retired General Count Ferdinand von Zeppelin who led the way, and, by November 1907, work had begun on a mammoth 136-metre (446-foot) airship, the LZ4. It made its maiden flight on 20 June the following year, and on 1 July 1908 covered 385 kilometres (240 miles) in twelve hours. This was, for

the day, an almost unimaginable distance. With French aviation still in its infancy and British aviation barely begun, Germany decided to focus much of her energy – and finances – into these dirigibles, as such airships were known.

In Britain, however, it was the Royal Navy that was the senior service. It was naval power that had helped her to rule the waves for a century and to become the centre of a vast trading network that included not only the biggest empire the world had ever known, but also trading posts and business assets in every corner of the globe. As Germany had begun building a modern navy, so Britain had responded with massive new battleships called dreadnoughts which had not only armour thicker than anything that had come before, but also mammoth guns. Powerful and state-of-the-art they may have been, but they were also incredibly expensive and

Above / LZ4, the fourth Zeppelin to be built, was destroyed in August 1908 when its hydrogen ignited during an attempt at a 24-hour endurance flight.

Opposite / First successful flight of the Wright Flyer, by the Wright brothers.

it was into these new-generation battleships that much of Britain's defence spending was directed.

In any case, the British government believed that with aviation very much in its infancy, there was a strong case for not investing too heavily in it at this early stage. Instead, they chose to watch carefully and rely on private investors, inventors and adventurers to lead the way. This was not an unreasonable approach. After all, Santos-Dumont's pioneering aircraft flight in Paris managed just 60 metres (65 yards) and never got more than five metres (16 feet) off the ground. Exciting though this was, it remained a long way from becoming a practical machine or weapon of modern warfare.

Furthermore, private enterprise did plough a first furrow in Britain. Lord Northcliffe, the owner of the *Daily Mail* newspaper, was concerned that Britain was being left behind in the aviation race and so, just after the first French flight, offered a prize of £10,000 for the first person to fly from London to Manchester. This was an enormous sum and although his suggestion was widely ridiculed at the time, it had the desired effect of encouraging private pioneers. One such man was Lieutenant-Colonel John Capper of the Royal Engineers, who commanded the Army's Balloon Section based at Farnborough in Hampshire. In 1904, he had made a trip to the United States and had visited the Wright brothers, returning convinced that aviation had a huge future. Another was Lieutenant John Dunne, a fellow officer in the Engineers, and together they began work on creating their own aircraft. So confident was Capper, that he twice urged the government to turn down an offer from the Wright brothers to sell their airplane to the War Office, which the British government duly did. The Wrights' early aircraft required constant manipulation to keep it airborne, whereas Capper and Dunne believed theirs would be inherently stable, so that it could become an invaluable reconnaissance tool. The more stable the machine, the easier it would be to spot enemy troop movements from the sky and even take photographs of them.

Capper and Dunne were certainly right in principle, but their machine was a failure and the miserly tap of financial support from the government was immediately switched off. In fact, by April 1909, the British government had invested just £5,000 in aeronautics, whereas in France that figure was nearly £50,000 and in Germany some £400,000. Nonetheless, the government was finally poised to invest in aviation. Back in 1908, Wilbur Wright had visited Europe and wowed crowds with displays of flying his latest machine. This then prompted Northcliffe to offer £1,000 to anyone who could fly the English Channel. On 25 July 1909, the Frenchman Louis Blériot, achieved the feat, flying from Calais to Dover in just 37 minutes. Six years after that first flight at Kitty Hawk and three after the first European flight, the potential of powered aircraft was becoming startlingly obvious. While Blériot's achievement was widely viewed with excitement, there were many in Britain who were beginning to feel unnerved, and those included Britain's military and political leaders. Suddenly, the English Channel did not appear to be quite the barrier it had always been.

Over the following four years, the government's earlier laid-back approach to aeronautical development was replaced by a growing sense of urgency. The Balloon Factory at Farnborough was renamed the Army Aircraft Factory and then became the Royal Aircraft Factory a year later; the King's stamp of approval was a notable signal that attitudes had changed. Dynamic young engineers were recruited to Farnborough, not least Geoffrey de Havilland, who quickly made a mark with his work on the first aircraft to be developed, the "BE" – or Blériot Experimental – and the "FE" – the Farman Experimental.

Wider public interest in aviation began to grow too. Aerodromes sprung up: at Eastchurch; at the motor racing circuit at Brooklands; and at Hendon in North London. The first hangars were built at Larkhill near Stonehenge on Salisbury Plain and this became the centre of aviation for the Army. Then, in February 1911, the Air Battalion was created within the Royal Engineers. Pilots were

drawn from all branches of the Army, but first had to gain their pilot's licence from the new Royal Aero Club.

Events were now moving fast. More pilots were coming forward, new machines were being developed, greater investment – private and public – was being thrown into this exciting new world of aviation, and in April 1912, the Air Battalion was superseded by the creation of the Royal Flying Corps, which was to have both a Naval and Military Wing and a Central Flying School at Upavon near Larkhill.

Yet although the RFC – as it quickly became known – had been created as a service of both the Army and Navy, in reality its organic growth turned out somewhat differently. Already, naval flying had begun to develop its own identity and organization, with a Naval Flying School starting at Eastchurch in 1910 and a greater dependence on private aircraft manufacturers rather than those emerging from Farnborough. By the eve of the First World War in July 1914, the Naval Wing had been cast aside in favour of the

Above / French aviator Louis Blériot's cross-Channel flight in 1909 demonstrated the possibilities of aviation, but also highlighted Britain's vulnerability.

Opposite, above / Blériot's monoplane is captured just as it passes the White Cliffs of Dover at the end of his pioneering flight across the English Channel.

Opposite, below / Geoffrey de Havilland, Britain's leading early aircraft designer, sits in the cockpit of a BE3, one of the first models he designed.

Royal Naval Air Service, or RNAS. The RFC had become purely a part of the Army.

When war was declared between Britain and Germany on 4 August 1914, the RFC had seven squadrons, although 1 Squadron had only just begun switching from airships to airplanes and 6 and 7 Squadrons were yet to reach full strength. What is more, they were heading to war in a range of aircraft built in England as well as in France; some had been constructed at Farnborough and others by private manufacturers. All of them looked flimsy and vulnerable and were just that. Flying was dangerous, and that was without being shot at. In fact, the RFC's first casualties were a pilot and mechanic who crashed shortly after take-off on 12 August, while en route to France.

Above / The group of officers forming the first course at the Central Flying School at Upavon includes future Marshal of the RAF Hugh Trenchard (centre row, far right).

Right / The brass cap badge of the Royal Flying Corps bears its monogram enclosed by laurel leaves and topped by a crown.

The four long years of the First World War were a period of astonishing growth and change for British aviation and for aviation generally. Fortunes ebbed and flowed as one side took the technological and production lead, then the other caught up and overtook it until the advantage switched yet again. Certainly, the RFC that went to war in August 1914 looked very different from the massively increased Royal Air Force that was born in April 1918. In fact, despite the new official aircraft from Farnborough, those seven squadrons were equipped with a staggering array of aircraft, including more than twenty French varieties, as well as those made by independent British manufacturers. What they did all have in common was an extremely fragile appearance. None looked as though they would withstand the rigours of aerial combat.

At this time, few thought of aircraft as offensive weapons. Rather, they were designed to provide invaluable aerial reconnaissance, and during the early months of the war aircraft were unarmed and pilots and observers fought each other with pistols and rifles. In any case, a lot of the reconnaissance work was pretty hit and miss. The

RFC's first reconnaissance flight was carried out by Philip Joubert de la Ferté and Gilbert Mapplebeck on 19 August. They did not see a huge amount of troop movements, but did fly over a large town, which only on their return did they learn had been Brussels. A lack of decent maps certainly didn't help.

The Eastchurch Squadron of the RNAS soon joined the RFC across the Channel, as it was felt they would be more use based in Dunkirk than in England, where there was still no sign of the much-feared German Zeppelins. Under the command of the colourful and tenacious Captain Charles Rumney Samson, they began using wireless telegraphy – radio – to report back on enemy troop movements and then started sending aircraft on bombing operations. On 8 October, Flight Lieutenant Reginald Marix and Squadron Commander Spenser Grey flew around 320 kilometres (almost 200 miles), first to Cologne, where Grey dropped bombs on the railway station, and then Marix, in the privately

Below / First Lord of the Admiralty Winston Churchill stands in front of a Short S.38 biplane during a May 1914 inspection of the Naval Flying School at Eastchurch.

manufactured Sopwith Tabloid, swooped low over Düsseldorf and managed to hand drop his bombs directly on to the Zeppelin shed roof. The entire building erupted into flames and a Zeppelin inside was destroyed. Marix was lucky to escape unscathed, but his Sopwith was peppered with ground fire, and about 30 kilometres (twenty miles) from the squadron airfield he was forced to land and abandon his plane, completing his journey on a borrowed bicycle. It was the first-ever bombing raid on Germany.

In December, Commander Samson carried out the first-ever night bombing raid, while on Christmas Day, the Royal Naval Air Service rounded off the year by launching the first seaborne air attack lowering a seaplane from a ship, which then took off and attacked Cuxhaven naval base. The RNAS were certainly showing the way when it came to taking air power on to the offensive.

As 1914 gave way to 1915, so the RFC and RNAS began spreading their wings. In April 1915, Lieutenant William Rhodes-Moorhouse won the first aerial Victoria Cross when he attacked a railway junction at Courtrai in an effort to delay German reinforcements heading to the front. Wounded by ground machine-gun fire en route, he pressed on; sweeping down low, he hand-dropped his bombs successfully from just 90 metres (300 feet), largely destroying the junction in the process, but was also severely wounded by groundfire. Somehow, Rhodes-Moorhouse managed to nurse his plane back, but died the following day. His exploits were immortalized by Errol Flynn in the 1938 film *The Dawn Patrol*. In 1916, the first aircraft carrier, HMS *Ark Royal*, set sail for Gallipoli with Commander Samson and six seaplanes, while the RFC was also operating aircraft in Egypt and resupplying troops besieged in Kut in Mesopotamia (modern-day Iraq).

The RNAS were not long in getting their own VC. In June, Lieutenant Reginald Warneford managed to drop a bomb on a Zeppelin while it was in flight. The blast of the exploding airship dislodged the fuel pipe on his plane and so he landed his aircraft behind enemy lines, managed to repair it and then flew back to base. Warneford was awarded the VC in a matter of hours, and by personal telegram from the King, but he had little chance to bask in the glory. Just a few weeks later, he was dead – killed in a flying accident.

Opposite / Zeppelin LZ37 plunges to the ground in flames after being bombed by Lieutenant Reginald Warneford; it was the first Zeppelin destroyed in air-to-air combat.

Above Right/ An early RFC volunteer Lieutenant William Rhodes-Moorhouse was the first airman to receive a VC, for the raid on Courtrai, during which he suffered a fatal wound.

Right / Philip Joubert de la Ferté (left) carried out the RFC's very first reconnaissance mission in August 1914 and rose to become Inspector-General of the RAF in 1943.

That summer, the German flyers took the lead. Air fighting had begun to change dramatically. No longer were aircrew taking pot-shots at each other with rifles. Machine-guns had begun to be mounted onto aircraft and fighter planes, or "scouts", were emerging to specifically protect their own craft and hunt down those enemy aircraft carrying out reconnaissance work. The first "aces" had emerged – dashing pilots with five or more "kills" to their name – and in Germany, men like Max Immelmann and Oswald Boelcke were fêted in Germany as pin-up heroes. They were also pioneering new tactics and means of manoeuvring into position to get a clear shot at an enemy plane. Helping them was new technology: interruptor gear, as it was known, although synchronized firing was a more accurate description.

The French ace, Roland Garros, had developed a propeller protected with a steel deflector that ensured he could fire a machine-gun through a rotating prop without causing it to splinter. The new German Fokker Eindeckers, however, fired in a synchronized fashion that ensured the bullets never hit the propeller blades. Anthony Fokker was, in fact, a Dutch engineer who had offered his services to both the French and British, but had been turned down. Instead, he was employed making rather good aircraft for Germany and now with machine-guns that could fire straight ahead. This gave the Germans a lethal advantage.

Top / A German officer poses in front of his Fokker monoplane. The new generation of aircraft armed with synchronized machine-guns gave German aviators a huge advantage.

Above / Roland Garros turned away from a career in music to become one of France's leading flying aces, aided by his use of a device that allowed him to fire a machine gun through a rotating propeller.

ACES HIGH

Britain's war leaders were reluctant to make heroes of individuals, but the media had no such qualms about elevating the gallant knights of the sky, whose courage, dash and daring in this most modern form of warfare seemed impossibly glamorous. Men like Reginald Warneford and William Leefe Robinson, the Zeppelin destroyers, were early idols, as was Major Lanoe Hawker, one of the RFC's first aces. He was eventually killed by Manfred von Richthofen, the "Red Baron", after a long, twenty-minute dog fight in which both men twisted, turned and looped about the sky in an effort to get on the tail of the other.

Another to become famous throughout Britain was Captain Albert Ball. Quiet, shy and devout on the ground, he turned into a reckless killer once airborne, and officially shot down at least 44 enemy aircraft by relentlessly attacking and firing at close quarters. Another VC winner, he was eventually shot down in May 1917 after becoming disoriented in thick cloud. A later ace was James McCudden, VC, who had gone to France in 1914 as a member of the ground crew, but who learned to fly and became a cool tactician and superb technical flyer. He, too, was never shot down, but died in an inexplicable accident in July 1918.

Perhaps the greatest British ace of them all, however, was Major Edward "Mick" Mannock, who was credited with 73 victories but probably had many more. A virulent hater of Germans, he was a superbly instinctive pilot, a brilliant shot and one who – unlike Ball – fought with calculated cunning. He was also the most decorated of British pilots, with a Military Cross and Bar, the Distinguished Service Cross and two Bars, and a Victoria Cross. He, too, was killed in July 1918.

All these men were brilliant, curiously good looking and dashing, and seemed to symbolize the glamour of the air war. The reality was, of course, nothing of the sort, and all four men flew far longer than they should have been allowed, and suffered more than their fair share of shattered nerves. They all died woefully young and tragically, as did so many other airmen.

Top / Nottingham-born Albert Ball became Britain's leading ace until his death in a May 1917 dogfight with a German squadron including the Red Baron's brother, Lothar von Richthofen.

Above / Major "Mick" Mannock served in the Royal Engineers before transferring to the RFC in 1916. He made his first confirmed kill in July 1917, the first of at least 73 victories.

Although Immelmann was killed in June 1916, the summer of 1915 was known as the "Fokker Scourge", and by August it was clear the RFC was in danger of being shot out of the sky altogether. Something had to be done.

In August, Colonel Hugh Trenchard was promoted and given overall charge of the RFC in France, taking over from Major-General David Henderson. He was determined to build up a superiority in numbers over the Germans, and recognized that the RFC urgently needed their own specialist fighter plane and a new improved breed of two-seater bomber and reconnaissance aircraft; two-seaters were essential for these latter roles because, while the pilot flew the plane, the second man had to take photographs or drop bombs. The British also needed to develop their own system of firing synchronization. Another Trenchard priority was to supply squadrons with just one type of aircraft each rather than the hotch-potch of differing varieties that was currently the norm. One of the problems was that new aircraft were supposed to come only from

Above / Wrecked houses at St Peter's Plain, Yarmouth, after a Zeppelin bombing raid in January 1915.

Opposite / A recognition guide produced by the British government in 1915 to help the public and military observers distinguish between friendly planes and possible German bombers.

the Royal Aircraft Factory and other officially sanctioned firms. More and more BEs and FEs were being produced, which were then sent to France and promptly shot down. It was scandalous and too many RFC airmen were losing their lives as a result. In contrast, the RNAS was using aircraft produced by independent firms such as Sopwith, Avro and now de Havilland (who had left Farnborough and set up on his own). These aircraft appeared to be manifestly superior to those in use by the RFC.

Meanwhile, the first Zeppelins started to appear over Britain. The first ever raid was on Great Yarmouth on the East Anglian coast in January 1915. The damage was very light and just four people were killed, but it demonstrated Britain was no longer isolated by the sea. More raids followed, including the first on London on 31 May the same year. These raids rarely inflicted much damage, but in Britain the reaction was one of horror and outrage that for the first time enemy bombs could be dropped on to home soil from the air.

Zeppelins certainly looked menacing, but their bomb load was small, they were extremely unwieldy, were easily sent off course by high winds and were filled with highly flammable gas. Nonetheless, they were proving hard to shoot down – initially, at any rate. Anti-aircraft guns were brought in to protect the capital and elsewhere, and, by February 1916, home defence had been handed to the

PUBLIC WARNING

The public are advised to familiarise themselves with the appearance of British and German Airships and Aeroplanes, so that they may not be alarmed by British aircraft, and may take shelter if German aircraft appear. **Should hostile aircraft be seen,** take shelter **immediately** in the nearest available house, preferably in the basement, and remain there until the aircraft have left the vicinity: do not stand about in crowds **and do not touch unexploded bombs.**

In the event of **HOSTILE** aircraft being seen in country districts, the nearest Naval, Military or Police Authorities should, if possible, be advised immediately by Telephone of the TIME OF APPEARANCE, the DIRECTION OF FLIGHT, **and whether the aircraft is an Airship or an Aeroplane.**

GERMAN

AIRSHIPS

BRITISH

AIRSHIPS

Note specially the shape of the Airships and the position of the passenger cars

ZEPPELIN

SCHUTTE – LANZ

PARSEVAL

H.M.A. ASTRA TORRES

H.M.A. BETA

H.M.A. ETA

H.M.A. PARSEVAL

Note specially the sloped-back wings of the German Aeroplanes

AEROPLANES

STAHLTAUBE MONOPLANE

RUMPLER TAUBE MONOPLANE

AVIATIK BIPLANE

ALBATROSS BIPLANE

D.F.W. BIPLANE

AEROPLANES

BRISTOL BIPLANE

BRISTOL BIPLANE

AVRO BIPLANE

AVRO BIPLANE

SHORT BIPLANE

B.E. BIPLANE

SOPWITH TRACTOR BIPLANE

H. FARMAN BIPLANE

SOPWITH TRACTOR BIPLANE

LONDON:
PRINTED UNDER THE AUTHORITY OF HIS MAJESTY'S STATIONERY OFFICE.
By SIR JOSEPH CAUSTON & SONS, LIMITED, 9, Eastcheap, E.C.
To be purchased, either directly or through any Bookseller, from WYMAN & SONS, LIMITED, 29, Breams Buildings, Fetter Lane, E.C.
and 54, St. Mary Street, Cardiff; or H.M. STATIONERY OFFICE (Scottish Branch), 23, Forth Street, Edinburgh; or E. PONSONBY,
LIMITED, 116, Grafton Street, Dublin; or from the Agencies in the British Colonies and Dependencies, The United States of
America, the Continent of Europe and Abroad of T. FISHER UNWIN, London, W.C.
1915.

PRICE TWOPENCE

COPYRIGHT
Sir Joseph Causton
& Sons, Ltd.
London.

BOOM
TRENCHARD

General Sir Hugh Trenchard was a towering figure in the early days of the RFC and the embryonic Royal Air Force. Known as "Boom" because of his resonant voice, he joined the Army as a young man, and had an indefatigable drive and energy despite losing a lung when badly wounded during the Anglo-Boer War. No-nonsense, bullish and something of a martinet, he learned to fly in 1912 and went on to become second-in-command of the Central Flying School, before taking command of the RFC in France in August 1915.

Trenchard not only moved to homogenize squadrons and then wings of squadrons, he also insisted on an almost entirely offensive air strategy to marry with the offensive operations the British were launching on the ground. This meant taking the attack to the Germans, which in turn called for crossing over enemy lines and into hostile territory. Clearly, effective reconnaissance was a huge asset before any offensive on the ground, but reconnaissance aircraft were vulnerable, and so scouts were sent ahead to form a protective barrier. The German tactic was to wait and then to pounce with the advantage of height and the sun behind them. Trenchard also insisted on bombing raids deep behind enemy lines, another extremely dangerous tactic. The trouble was, not only did RFC pilots and aircrew risk being attacked by German fighters, they also faced attack from the ground, something they could have avoided if they had remained behind their own lines. Both the front itself and key supply dumps and

Above / Sir Hugh Trenchard rose from commander of the Central Flying School to become Chief of the Air Staff and a redoubtable defender of military aviation's importance.

installations further back were bristling with "Archie", as anti-aircraft fire was known. As a result, casualties among the RFC were very high.

Trenchard was intractable, but his zealous pursuit of offensive operations did mean that the RAF, when it came into being, was already imbued with an aggressive spirit – one it would carry into the final months of the war and beyond.

RFC, while the RNAS was to continue to try and prevent the Zeppelins ever reaching Britain in the first place. Incredibly, it was not until early September 1916 that the first German airship was shot down over Britain. Lieutenant William Leefe Robinson, using incendiary ammunition in his machine-guns, caused the airship to catch fire and, burning, it plummeted to the ground in full view of thousands of screaming, cheering spectators. Leefe Robinson swiftly became a national hero and was also awarded the VC.

More victories over the Zeppelins swiftly followed, by which time VCs were not being dished out quite so readily. By the end of 1916, there had been a total of 51 raids, which had killed around 500 civilians. However, over 50 Zeppelins had been lost. Facing increased fighter attacks and greater ground fire, they were forced higher, which in turn meant they were even less accurate and more susceptible to wind. A number were simply blown off course and vanished forever. These were unsustainable losses and the raids were stopped. The Zeppelin Terror was over – for a time, at least.

Above / Lieutenant William Leefe Robinson in the BE2C in which he downed the first Zeppelin shot down over British soil.

Left / Leefe Robinson's feat in bringing down a German Zeppelin made him one of the very few men to earn a Victoria Cross for action in Britain itself.

Meanwhile, by the summer of 1916, the RFC squadrons had become more homogenized into specialist reconnaissance, bombing and fighter squadrons, equipped with new improved FE2s and, at last, some independently produced aircraft too. At the Battle of the Somme in July, the RFC had 47 squadrons and well over 400 aircraft, many more than the Germans. One of Trenchard's main aims had been achieved and this superiority in numbers was something the RFC would maintain until the end of the war.

Although the RFC had wrested control of the skies over the battlefield by the launch of the Somme offensive, their ascendancy was brief, as by the autumn new superior German aircraft types had appeared. The Albatross D.I and Halberstadt scouts were significantly superior to the latest BE2s and FE2s. A new Bristol Scout had come into service, but not in significant numbers, while the Albatross was further upgraded with the introduction of the D.II, and these scouts formed into larger fighter formations known as circuses.

Over Arras in April 1917, the young men of the RFC were slaughtered. In 1916, the life expectancy of a pilot in the RFC was around ten weeks, but by the spring of 1917, it had fallen to less than two months. The numbers of new recruits could not keep up with losses and so training was cut and more and more young men were being sent over to France woefully underprepared. The previous summer, the MP Noel Pemberton Billing had accused the government of not just killing its young aircrew but murdering them. This accusation had been dismissed but the young men of the RFC were certainly being shot down in even greater droves during what became known as "Bloody April".

The rot was largely stopped with the end of the Arras offensive in mid-June 1917 and, at long last, the introduction of new much-improved aircraft such as the SE5a and the famous Sopwith Camel, which were more than a match for the latest German models and were equipped with synchronized machine-guns.

By this time, however, Britain was once more under attack from the air, not from Zeppelins but from the latest German weapon: the monstrous Gotha G.IV. With a wingspan of 23.8 metres (78 feet), two engines and capable of carrying seven 50 kg (110 lb) and six 12.5 kg (27.5 lb) bombs, it was a demonstration

Below / A German Albatross C.III over the Western Front. Introduced in early 1916, it became the most common German reconnaissance plane and was also used as a light bomber.

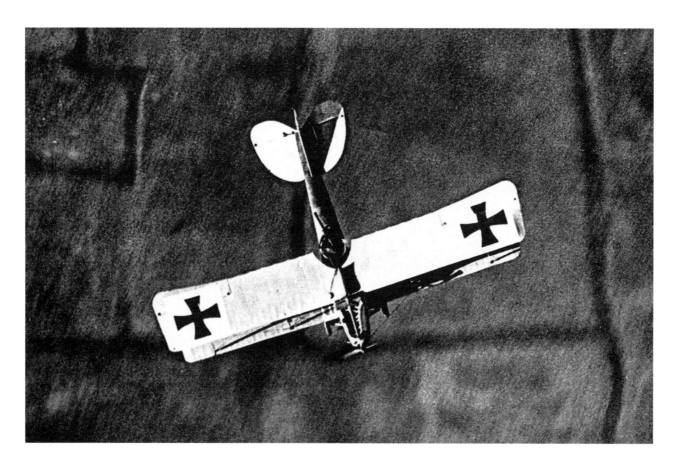

DAWN PATROL

"We got lost in a terrific ball of cloud and come out well over Hun lines," scribbled Lieutenant Peter Wilson on 26 March 1918. "We are heavily archied and I am very windy as I am only in pyjamas under flying kit. Keep a sharp look out for Hun scouts." The next day he added, "New Pilot killed in crash." Wilson's diary was typical of the casual insouciance that was very much part of the culture of the RFC, but flying at all, let alone in combat, required nerves of steel. British flyers were refused parachutes in case these made them too eager to bale out, and, for many, a clean bullet through the heart or head was the best they could hope for. More often, they burned to death or suffered a terrifying plummet out of control, ending in a fireball on the ground.

Despite this, there was never a shortage of volunteers. Flying was not only glamorous, it was also seen as exciting and preferable to the mud and dust of the trenches. In between flying, aircrew lived in comfort, with plenty of food, drink, comfortable beds and billets sited well back from the front lines. Camaraderie amongst flyers was intense, but many suffered from the extremes of brief moments of searing terror thousands of feet up in the sky followed by the sudden return to calm comfort back on the ground. It was the contrast between the repeated sight of burning men and machines and mangled aircraft and bodies and the easy-going atmosphere of the mess that shattered the nerves of many young flyers.

Training was rarely as good as it might have been, and by early 1917, new pilots were arriving with only a few hours of flying in their logbooks. Many never even got to the front: of the 14,000 airmen who lost their lives in the First World War, more than 8,000 were killed in training or in accidents. Most were young, keen and idealistic, and really were lambs to the slaughter. The key was to survive training, then the first few weeks, build up flying hours, and learn a few tricks; then the chances of survival rose dramatically. Luck always played a huge part, however, and the simple equation was this: the longer a pilot survived, the greater was the chance of his luck eventually running out. It was also true that the longer a pilot flew, the greater the chances were of his succumbing to what is now understood to be combat stress. There were usually tell-tale signs, but tragically, all too often these signs were not understood and acted upon.

of just how far air power had developed in only a few years. And unlike the airships, the Gothas really were a potent force. Flying in formation, sometimes of more than twenty machines, they began terrorizing Britain from May 1917. Civilian deaths rose dramatically, as did the amount of damage. The bombing often appeared indiscriminate. On 13 June, bombs fell on London's East End, including on a school in Poplar, killing eighteen children and wounding many more.

The outcry was enormous and anti-German sentiment rose dramatically, but so too did the level of defiance of the British people. Home defence was rapidly increased with more anti-aircraft guns, searchlights and fighter squadrons being put into place. By the autumn of 1917, the Gothas were starting to be shot down in droves – five were destroyed out of 21 in one raid alone at the end of October – and although the raids continued into 1918, they never succeeded in bringing Britain to her knees. What is more, the Germans, now struggling badly economically under the pressures of sustaining their war effort, were unable to make good the losses from the raids. In May 1918, raids on Britain were brought to an abrupt end and the remaining Gothas were turned to operations over the front instead.

The Gotha raids did, though, play a highly significant part in the development of British air power. Throughout the war, there had been various boards set up to investigate the shortcomings of the RFC, and back in May 1916 the Air Board had been established under Lord Curzon, before Lord Cowdray had taken over in December the same year. Its task was to oversee the production and supply of aircraft to both the RFC and RNAS, but once the Gotha raids began, however, it was clear to even the most blinkered of government mandarins that Britain's air arms also needed greater support in terms of deployment, tactics and strategy. That summer of 1917, as the Gothas thundered over, the South African

An artist's impression of the first German Gotha bomber shot down over British soil, on 28 January 1918. The British pilots, Captain George Hackwill and Lieutenant Charles Banks, flying Camels, both received Military Crosses.

Lieutenant-General Jan Smuts, once Britain's enemy during the Anglo-Boer War, but now a loyal subject, was invited by the Prime Minister to join the British war cabinet and the War Policy Committee and was given a specific brief to review Britain's air power. Smuts was a highly respected and intelligent man with no axe to grind, and it was recognized he could offer an objective and largely independent judgement. He recommended there be one Air Ministry and one combined air force, suggestions that were passed into law by the Air Force (Constitution) Act on 29 November 1917.

Putting this into practice took a little time, however, and progress towards creating a unified national air force was hampered by extraordinary ructions between the key players. The Air Council was formed on 3 January with Lord Rothermere proposed as the first head of the Air Ministry, a post many had assumed would go to Lord Cowdray. However, Cowdray had fallen foul of the Prime Minister, David Lloyd George. General Trenchard was recalled from France to become Chief of the Air Staff, the military head of the new RAF. However, relations between Rothermere and Trenchard were beset from the outset by differences of opinion and mutual distrust. It was a disaster, and Trenchard submitted his resignation in March just two weeks before the Royal Air Force formally came into being. It was not a promising start, but nonetheless, on 1 April 1918, both the RFC and RNAS ceased to exist and the Royal Air Force was born.

Top / The Sopwith Camels parked in this snowy field in northern France include that of Flight Commander G.W. Price, a RNAS ace who scored a dozen victories before being shot down in February 1918.

Above / The appointment of the newspaper baron Lord Rothermere as the first Air Minister in 1918 was bedevilled by clashes with Hugh Trenchard, his military counterpart.

Opposite / General Jan Smuts' report recommending the formation of the RAF in August 1917.

REPORT BY GENERAL SMUTS ON AIR ORGANISATION

AND THE DIRECTION OF AERIAL OPERATIONS.

August, 1917.

Our first report dealt with the defence of the London area against air raids.

We proceed to deal in this report with the second term of reference: the Air organisation generally and the Direction of Air Operations. For the considerations which will appear in the course of this report we consider the early settlement of this matter of vital importance to the successful prosecution of the war. The three most important questions which press for an early answer are:-

1. Shall there be instituted a real Air Ministry responsible for all Air Organisation and operations.

2. Shall there be constituted a unified Air Service embracing both the present R.N.A.S. and R.F.C? And if this second question is answered in the affirmative, the third question arises:-

3. How shall the relations of the new Air Service to the Navy and the Army be determined so that the functions at present discharged for them by the R.N.A.S. and R.F.C. respectively shall continue to be efficiently performed by the new Air Service?

Opposite / Mechanics service an Airco DH4, a US-built version of which formed the mainstay of the US Air Force in Europe during the First World War and which remained in service until 1932.

2

THE
EARLY
YEARS
OF THE
RAF

There were no great fanfares when the RAF came into being. For most of its personnel, one day they were wearing khaki and the next were being issued new Royal Air Force blue uniforms. Otherwise, it was business as usual and that, in the spring of 1918 – along the Western Front at any rate – was all about supporting the troops on the ground. Back in March, the Germans had launched a major offensive and both the RFC and RNAS had been busy carrying out reconnaissance work, but also trying to shoot up and drop bombs on the Germans swarming across open ground.

The German attack had been directed towards Amiens and the British in that sector had been forced to hurriedly give up a great deal of land. More than three years of static trench warfare had suddenly become mobile again. On 9 April, the Germans launched the second phase of their offensive and again broke through in a number of places. Flying Officer Peter Wilson was a pilot in 7 Squadron flying twin-seater RE8 reconnaissance bombers. On 10 April, he was up at 4 a.m. and soon after was on a bombing mission with five other aircraft, flying low over countryside that just a couple of days earlier had been well within their lines, but was now overrun by the enemy.

Flying at just over 60 metres (200 feet), they saw a number of farms burning and came under heavy machine-gun attack from the ground. One aircraft caught fire and dived, then they were over Armentières and watching the French forces evacuating. As they returned to base, Wilson saw another plane explode on landing.

"Ours the <u>only</u> machine to return from the initial six sent out to check German advance," he noted in his diary. Later that day, he was flying again, dropping bombs on a village they had been billeted in back in 1915, and then strafing troops on the ground. "Flying low," he jotted, "I fire 400 rounds at masses of Hun troops." That sortie – as individual flights were known – he survived, too, although others were not so lucky. It had been a brutal day for 7 Squadron. The RAF was playing a vital role in slowing the German advance, but the cost was horrific: daily casualties of 30 percent, which meant a statistical turnover of a squadron's flying personnel in just four days.

Fortunately, by 12 April the German advance had been checked. The offensive had been the enemy's last throw of the dice on the ground and in the air. As if to underline the point, the great German ace, Manfred von Richthofen, the "Red Baron", was killed on 21 April. "I hope he roasted the whole way down!" muttered Major Mick Mannock on hearing the news. By this time, Mannock was fast becoming the RAF's leading ace and symbolized the new ascendancy of the Allied air forces. For all the horrendous casualties of its first two weeks, the RAF was growing rapidly – as was the French air component – and the Germans, at this stage of the war, became increasingly outnumbered.

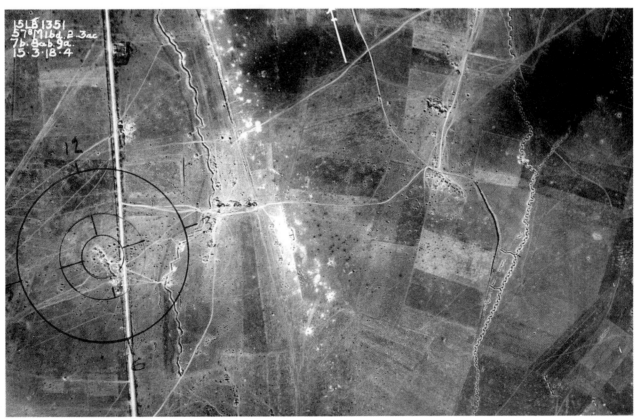

Opposite / Members of 12 Squadron with their RE8 biplanes, a reconnaissance type introduced in mid-1916. Heavy early casualties and a reputation for being underpowered made it unpopular.

Top / Frederick Sykes was appointed Chief of the Air Staff on 14 April 1918 following Trenchard's resignation. He played a key role in consolidating the new service.

Above / An early RAF reconnaissance photo. Reconnaissance was one of the first military roles for aircraft, providing vital intelligence about enemy deployments and defences.

Crew members stand in front of a Handley Page O/400 bomber. The first purpose-designed strategic bomber, it saw widespread use in the closing stages of the war.

Meanwhile, the RAF was developing what it called a "strategic" bombing wing – that is, a bombing force that could operate independently. The RNAS had begun this process and Commander Samson had been given command of 3 Wing for the purpose. Then the RFC had followed suit, and by February 1918 had formed VIII Brigade, which amongst its arsenal included the Handley Page 0/400, the first specifically designed strategic bomber. In June 1918, VIII Brigade was expanded further and became the Independent Bombing Force. Commanding this new bomber force within the RAF was none other than General Sir Hugh Trenchard, who had earlier resigned as the first head of the Air Force after repeatedly clashing with Lord Rothermere, the Air Minister. For the RAF – and its future development – this was one of the most significant appointments ever made, because although at this time Trenchard harboured some doubts about the value of his force's work, he went on to become a fervent advocate of bombing and the vital importance of strategic air power, and the seeds of those convictions occurred in these final months of the war, when his bomber force, flying at night and with no navigation aids, launched a number of raids on German industrial plants, including Saarbrucken on 17 September. In all, some 500 tons of bombs were dropped on Germany by the war's end, small beer compared with what was

to follow in the years to come. How successful the bombing raids were is debatable, but they did enough to set the RAF on a course from which it would not deviate for many years.

As the British went on the offensive, the key tasks facing the RAF were to both destroy the German Army Air Service and once more support the British Army's ground operations. By August 1918, the British had double the number of aircraft of the Germans and, taken together with the French, possessed a 4:1 numerical advantage. The Germans were also suffering from a lack of fuel. A knock-out blow would and should have been possible, but instead the RAF's punch was weakened by endless low-level – and costly – operations in direct support of the army. As a result, swirling dog fights continued over the front right up until the very end of the war.

Meanwhile, the former RNAS squadrons were now operating from aircraft carriers. The first custom-built deck carrier, HMS *Argus,* entered full service in October, but earlier, in July, seven Sopwith Camels had taken off from the converted battlecruiser HMS *Furious* and successfully raided the Zeppelin base at Tondern in northern Germany.

Below / Gloster Nightjars stand ready on the deck of HMS *Argus*, Britain's first custom-built aircraft carrier, during a 1922 expedition to Turkey.

The war finally ended on 11 November 1918. The British had gone to war with under a hundred planes and ended it with nearly 23,000 men in the new Royal Air Force – an extraordinary growth by any standards. However, more than four years of aerial warfare came at the terrible cost of 16,500 casualties, of which 6,166 had been killed in action. The end of the war also left the new third service in a vulnerable position. Most of the nearly 300,000 personnel in the RAF were now demobbed. With all combatant nations vowing not to go to war on such scale again and with Britain nursing crippling war debts, there were plenty who were suggesting the newest service should be scrapped.

In 1919 – the year of the Paris Peace Conference – squadron after squadron was disbanded. By the time the Treaty of Versailles was signed at the end of June that year, the RAF had shrunk to just 33 squadrons and its future appeared to be in some considerable doubt. David Lloyd George was re-elected as Prime Minister in December 1918 and appointed Winston Churchill as Secretary of State for both War and Air with the intention of disbanding the Air Ministry and with it the RAF. No-one doubted the importance of air power, but neither the Army nor the Royal Navy believed this required an independent air force.

Churchill, however, had been an enthusiastic advocate of air power for a decade, had even learned to fly himself, and instinctively believed in its future as an independent service. Nonetheless, in order to keep the wolves at bay, the post-war RAF had to prove its value swiftly, which was a very tall order indeed with a budget massively slashed to a fraction of those of the Army and Navy and with the Royal Navy – Britain's Senior Service, after all – desperate to see the air force kicked into touch.

A dynamic and strong Chief of the Air Staff was going to be essential, and, fortunately, Churchill chose wisely. Boom Trenchard had been out of a job since the end of the war and the disbandment of the Independent Bombing Force, but he had caught Churchill's attention, and when they met to discuss the matter had swiftly impressed him. In February 1919, Trenchard became Chief of the Air Staff once more, in what was to prove a vital appointment.

Trenchard has had his detractors. He had a reputation as a martinet, overbearing and stubbornly convinced that his own very single-minded vision for the future of the RAF must be the

Above / Winston Churchill stands with Hugh Trenchard, Chief of the Air Staff, during the Hendon Air Display in June 1920.

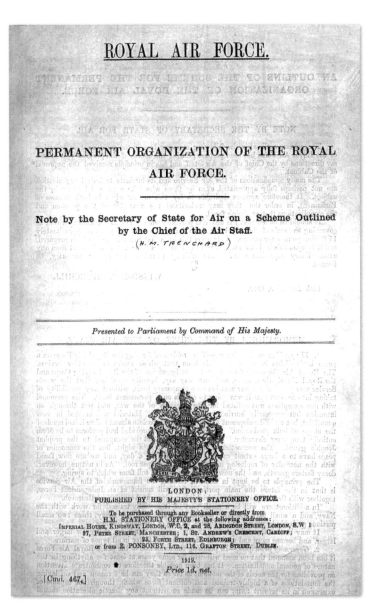

ROYAL AIR FORCE.

PERMANENT ORGANIZATION OF THE ROYAL AIR FORCE.

Note by the Secretary of State for Air on a Scheme Outlined by the Chief of the Air Staff.

(*H. M. TRENCHARD*)

Presented to Parliament by Command of His Majesty.

LONDON:
PUBLISHED BY HIS MAJESTY'S STATIONERY OFFICE.

To be purchased through any Bookseller or directly from
H.M. STATIONERY OFFICE at the following addresses:
IMPERIAL HOUSE, KINGSWAY, LONDON, W.C. 2, and 28, ABINGDON STREET, LONDON, S.W 1;
37, PETER STREET, MANCHESTER; 1, ST. ANDREW'S CRESCENT, CARDIFF;
23, FORTH STREET, EDINBURGH;
or from E. PONSONBY, LTD., 116, GRAFTON STREET, DUBLIN.

1919.

Price 1d. net.

[Cmd. 467.]

AN OUTLINE OF THE SCHEME FOR THE PERMANENT ORGANIZATION OF THE ROYAL AIR FORCE.

NOTE BY THE SECRETARY OF STATE FOR AIR.

The scheme outlined in the following memorandum on the permanent organization of the Royal Air Force has been prepared during the course of the present year under my directions by the Chief of the Air Staff, and has in principle received the approval of the Cabinet.

The many complications of the Air Service and its intricate technical organization are not perhaps fully appreciated, even by those who take a general interest in the subject. It therefore appears desirable to lay this memorandum in both Houses of Parliament, in order that they may understand the character of the problem and the complications that are being faced.

It should be added that the financial provision which the Cabinet have approved as governing the scale of the Royal Air Force during the next few years is approximately 15 million pounds per annum. It is upon this basis that this scheme has been prepared, and it is upon this basis that it is hoped the Estimates of next year will, apart from any extraordinary expenditure which the military situation may render necessary, be framed.

WINSTON S. CHURCHILL.

11th December, 1919.

MEMORANDUM BY THE CHIEF OF THE AIR STAFF.

1. *The problem confronting us.*—The problem of forming the Royal Air Force on a peace basis differs in many essentials from that which confronts the older services. The Royal Air Force was formed by the amalgamation of the Royal Flying Corps and the Royal Naval Air Service, and one may say, broadly speaking, that the whole Service was practically a war creation on a temporary basis, without any possibility of taking into account that it was going to remain on a permanent basis. The personnel with few exceptions was enlisted for the duration of the war, and put through an intensive but necessarily hurried course of training. Material was created in vast quantities, but rapid development often rendered it obsolete almost before it had reached the stage of bulk production. The accommodation provided had perforce to be of an entirely temporary character. The force may in fact be compared to the prophet Jonah's gourd. The necessities of war created it in a night, but the economies of peace have to a large extent caused it to wither in a day, and we are now faced with the necessity of replacing it with a plant of deeper root. As in nature, however, decay fosters growth, and the new plant has a fruitful soil from which to spring.

The principle to be kept in mind in forming the framework of the Air Service is that in the future the main portion of it will consist of an Independent Force, together with Service personnel required in carrying out Aeronautical Research.

In addition there will be a small part of it specially trained for work with the Navy, and a small part specially trained for work with the Army, these two small portions probably becoming, in the future, an arm of the older services.

It may be that the main portion, the Independent Air Force, will grow larger and larger, and become more and more the predominating factor in all types of warfare.

2. *Governing principles.*—In planning the formation of the peace Royal Air Force it has been assumed that no need will arise for some years at least for anything in the nature of general mobilization. It has been possible therefore to concentrate attention on providing for the needs of the moment as far as they can be foreseen and on laying the foundations of a highly-trained and efficient force which, though not capable of expansion in its present form, can be made so without any drastic alteration should

---

correct one. He also zealously believed that independent strategic air power had a vital role to play in Britain's future, and in this he was absolutely right. Few of Britain's military and political leaders understood this distinction or its implications, however, and there is no doubt that in fighting for his vision of the future RAF, Trenchard's forceful character, his drive, and his ambition for this newest of forces were unquestionably crucial to the growing strength and influence of the service. Combined with the equally forceful and bullish character of Churchill, the RAF now had two very powerful leaders to take it forward.

Nonetheless, the challenges they faced were enormous. With an annual budget of just £15 million, and with both the service chiefs of the Royal Navy and Army – as well as a number of leading politicians – deeply opposed to the existence of an independent air force, Trenchard knew he had to act swiftly and decisively. The

first task was to decide who should remain in the RAF, who should be released and which airfields and infrastructure to keep. He also had to consider how the RAF could best demonstrate its continued value and relevance.

Almost immediately, however, Trenchard discovered Churchill was beginning to waver in his support. The debilitating budgets and the Ten Year Rule, a policy which presumed Britain would be involved in no major conflict for at least a decade, made it increasingly hard to justify the expense of a third service. Matters came to a head with a blazing row between the two men, although tempers cooled enough for Churchill to ask Trenchard to write what amounted to a mission statement. This Trenchard did that very same day, 19 September 1919: a 3,000-word document making his case clearly both on long-term strategic reasons and explaining how the service's foundations could be secured within the very limited

budget already allocated. "Trenchard's Memorandum", as it became known, was submitted by Churchill as a formal government White Paper on 11 December 1919.

The 1920s and early 1930s have usually been portrayed as a time when the fledgling RAF was battling against shrinking budgets and struggling to compete with its big siblings, the Army and Royal Navy. The truth is that all three services suffered as a result of the drastic cuts to post-war defence budgets, but the Army and Navy suffered most, and the cuts actually served the purpose of the young RAF very well. This was because Trenchard was swiftly able to demonstrate that an independent strategic air force such as the

RAF could save Britain huge amounts in terms of manpower, time, and, by default, money.

His first aim was to establish firm foundations which would be very difficult to then unpick. It was essential, Trenchard knew, to create an elite and highly professional third arm, in which the highest standards of both training and performance were to be maintained. Again, these aims were easier to achieve in a small service. First, only the best of the very large wartime RAF were to be retained. Second, a brand-new officer cadet college at Cranwell in Lincolnshire, as well as a technical training college for non-commissioned officers at Halton in Buckinghamshire, were to be swiftly established.

Opposite / The government White Paper that set forward the strategy for the retention of an air force was heavily based on a September 1919 memorandum by Hugh Trenchard.

Above / Aircraft types on show for the 1932 RAF Hendon pageant include Fairey Hendons, Handley Page Heyfords and Westland Wallaces.

AIRCRAFT DESIGN AND TECHNOLOGY BETWEEN THE WARS

Another myth of the inter-war period is that Britain languished behind other nations in terms of air technology and that the low defence budgets meant low investment. This was not really the case: under Trenchard, the RAF spent considerably more on armaments than the Army. In 1924/25, the Navy spent £13 million, the RAF £6.9 million and the Army just £2.6 million. Eight years later, in 1932/33, those figures were £10.3 million by the Navy, £8.7 million by the RAF and just £1.8 million by the Army. In other words, RAF procurement was increasing, while that of the Army and Navy was falling.

With the end of the First World War, the Royal Aircraft Factory ceased designing and building aircraft and was once again renamed, this time as the Royal Aircraft Establishment, (RAE), with the brief to focus on research. By the mid-1920s, as much as 20 percent of the Air Ministry's entire budget was spent on research and development, and the RAE attracted many highly skilled and inventive scientists and engineers, including A.A. Griffith, who in 1926 discovered how to develop an effective axial flow compressor. This would be later harnessed with jet technology to give Britain a decisive lead in this area.

Meanwhile, most of the major independent aircraft manufacturers continued to design and build both for the RAF and for the burgeoning civil aviation industry. Sopwith was liquidated and renamed Hawker, but with Thomas Sopwith as its chairman. Other wartime firms, such as A.V. Roe, De Havilland, Gloster, Bristol, Handley Page and Short all continued to thrive and became increasingly profitable; into the mid-1930s, Britain was the world's leading exporter of aircraft.

Until the mid-1930s, most new aircraft designs were biplanes and to the uneducated eye looked much as they had at the end of the First World War. However, wing and airframe design, as well as the materials being used, were changing and evolving quite dramatically, as was engine design. One of the main aircraft engine manufacturers was

Above / The high performance of the Hawker Fury, introduced in 1931, never resulted in large production orders, as the RAF had already adopted the inferior Bristol Bulldog.

Opposite, top / Despite being slower than the Hawker Fury, the Bristol Bulldog became the most widespread fighter in RAF service in the 1920s, but never saw active combat.

Opposite, below / A Flying Flea aircraft is tested in a wind tunnel at the Royal Aircraft Establishment, the unit that produced many innovative designs for the RAF in the inter-war period and Second World War.

Rolls-Royce, who by the early 1930s was producing world-beating engines of greatly increased power, and, just as importantly, of vastly improved reliability and safety. While these technological changes were taking place, there were still very good reasons for persevering with biplanes rather than the monoplanes, which did not yet have a decisive edge. Monoplanes could offer greater speed, but there were other factors such as strength, weight and manoeuvrability, many of which were better answered with biplanes. Another factor was the lower landing speed of biplanes – and therefore shorter landing distance required – which was why they continued to be used on carriers even after the monoplane had in general won the ascendency after the mid-1930s.

At any rate, the RAF was certainly not lagging behind other countries in terms of aviation technology. On the contrary, both the RAF and the aviation industry as a whole had laid very firm foundations during the inter-war period, providing a firm base from which future air power could be greatly expanded.

A squadron of Hawker Harts. Designed by Sydney Camm, the Hart came into service as a fighter in 1930 and a variant was widely used as a training aircraft.

THE SCHNEIDER TROPHY

In 1912, during the first flush of pioneering excitement about the possibilities of aviation, French financier and air enthusiast Jacques Schneider announced a prize of around £1,000 – then a huge sum – for the winner of a race for seaplanes. The competition was to be held annually and it was agreed that should one particular club or manufacturer win it three times in five years, then they would retain the trophy for ever. The races were supervised by the Fédération Aéronautique Internationale and the national aero club of the host nation.

The Schneider Trophy was held just twice before the First World War – and the races were won by first the French, and then the British with a Sopwith Tabloid – but it was resurrected in 1919 when the Italians won, although they were later disqualified. Over the next few years, the trophy was won twice by the Italians, followed by the British and then twice by the Americans, the second time in 1925 with a Curtiss piloted by Jimmy Doolittle, later to become one of the leading US air commanders of the Second World War.

In 1927, three British manufacturers entered the race, which was won by Southampton-based Supermarine in their S5s, designed by R.J. Mitchell. By this time, the races were attracting huge international attention and were doing much to further a growing obsession with aviation among the leading nations of the world. The British participation was also backed by the government, which permitted RAF pilots to take part.

With Supermarine's first victory, however, it was decided to make the races biennial to allow manufacturers more time for development; as a result, the next races were not until 1929. Again, Supermarine won, with an average speed of 528.89 km/h (328.64 m.p.h.), unimaginably fast for the day and unthinkable just a few years before. The next races were to be held in 1931. By that time, with the country in the grip of the Great Depression, the British government had withdrawn support. Supermarine, however, secured private backing, and so R.J. Mitchell

Above / The S5, designed by R.J. Mitchell, brought Supermarine its first Schneider Trophy victory, in 1927. The winning pilot, Flight Lieutenant Sidney Webster topped 280 m.p.h.

Opposite / Although a squadron of DH9s were sent as bombers to support the British campaign in Somaliland in 1920, this aircraft was modified for use as an ambulance.

was able to develop a new, improved plane, the S6B. Like its predecessors, it was a low-wing monoplane and a seaplane. Ironically, despite the huge drag caused by the fixed floats, seaplanes were the fastest aircraft at the time because they could land at much higher speeds than land aircraft. Higher speeds once airborne were achieved through a combination of the engine, sleek aerodynamics and a high-speed wing. The gap in technology of the day lay not in creating a high-speed wing, but in designing one that could handle a wide range of different flying speeds. It was for this reason that seaplanes of this era could generally fly around 160 km/h (100 m.p.h.) faster than land-based aircraft.

The 1931 races were won for a third time by Mitchell's S6B with a race speed of 547.3 km/h (340.08 m.p.h.), although the plane also separately clocked a new world record of 611.5 km/h (380 m.p.h.) and later that year broke 643 km/h (400 m.p.h.). The British success in the Schneider Trophy was yet another reminder of the good shape the country's aviation was in by the 1930s.

Trenchard also made clear that the bulk of flying operations were to be overseas. Following the Treaty of Versailles, Britain's empire had grown yet further, which gave Trenchard the opportunity to test his theory that air power could save Britain both manpower and money. "The first duty of the Royal Air Force,' Churchill told Parliament as he delivered the White Paper, "is to garrison the British Empire."

Trenchard did not have to wait long to demonstrate the essential truth of his claim. In early 1920, the British attempted to quell a long-term insurrection in Somaliland in East Africa led by Mohammed Abdullah Hassan, a rebellion which had grown in strength during the First World War. A large and expensive land operation was planned to defeat Hassan and his followers, but Trenchard now offered air support as well. Just one squadron of twelve DH9s were sent from HMS *Ark Royal,* and began bombing Hassan's strongholds. Just a week after their first attack, the rebels were in full retreat. In less than three weeks it was all over and peace was restored. For just £77,000, the RAF had not only proved its value, but also saved the British Government from a potentially long and costly land campaign.

Buoyed by this success, Churchill now championed substituting RAF squadrons for large deployments of the army. By 1922, Parliament had given its support to the principle of air control of large swathes of the Middle East and the North-West Frontier between British India and Afghanistan. Although Trenchard still faced continued opposition from the Army and Royal Navy, the RAF had begun to make itself indispensable, while its body of supporters, from the Treasury to the Colonial Office, was increasing. Eight squadrons, for example, were sent to Iraq and played a vital role in quelling trouble there, too, which they were often able to do merely by demonstrating their power by flying over trouble spots. "The air force is a preventative against risings more than a means of putting them down," Trenchard told his Middle East Commander, Air Vice-Marshal John Salmond. "Concentration is the first essential. Continuous demonstration is the second essential. And when punishment is intended, the punishment must be severe, continuous, and even prolonged."

TAKORADI ROUTE

Trenchard had been keen to see the tentacles of the RAF spread to every corner of the British Empire and to use the air force to pioneer new air routes that could then both help with imperial defence and have the potential to be developed further by civilian operators. Bases were set up not just in the Middle East but also in India and even the Far East. Just as men had sailed and trekked across uncharted territory, now it was the turn of aviators, and it was seen as important that British pilots should be at the forefront of such expeditions.

A number of such pioneering expeditions were launched throughout the 1920s, but in 1925, the RAF in Egypt was given the challenge of linking British territories in West Africa to Egypt, then a British protectorate. Squadron Leader Arthur Coningham was to lead a team of three in their DH9s. They were to head from Helwan, just outside Cairo, to Kano in Nigeria, a journey of nearly 4,800 kilometres (3,000 miles). The challenges were

many. The aircraft were ageing, with engines that needed constant attention. They had no radio aids whatsoever, either on the ground or in the air, while the maps they had were primitive to say the least; for large parts of their journey they would simply have no features marked on the maps at all; they would be spending long hours in a cramped cockpit; and the landing strips they were to use along the way were rough and often short – even for an old de Havilland biplane.

Coningham, who was always known as "Mary", was an Australian raised in New Zealand. He had sailed for Britain at the outbreak of war and he eventually became a flying ace with the RFC. A natural leader and apparently fearless too, he was the ideal choice for the expedition and certainly seemed remarkably unfazed by the prospect.

The aircraft set off from Cairo on 27 October, heading south to Khartoum in Sudan, then turned west. Roughly in the centre of Africa, they touched down for a stop at El Fasher, in western Sudan. "The country from this point onwards," wrote Coningham later, "had never been traversed by aircraft." They flew at 900 metres (3,000 feet) and at just 130 km/h (80 m.p.h.) so as not to overly strain the engines. From that height, they could see as much as 240 kilometres (150 miles). There was little sense of speed at all and the temptation to open the throttle and hurry as the day's flying drew near to its end was enormous. Coningham resisted the urge, however, gently nursing his aircraft to the next stop.

But on 1 November, they did reach Kano in Nigeria, where a huge crowd of some 20,000 was waiting to greet them. In all, it had taken 36 hours and 50 minutes' flying time. After the celebrations and some rest, they then flew all the way back again, safely reaching Cairo on 19 November. The Air Ministry wasted no time in announcing the achievement, pointing out it was the first time aircraft had ever flown east-west across Africa. That same journey, it was pointed out, "by normal methods of rail, steamer, camel and bullock transport" would normally take six months.

Little did Coningham and his fellows know how important that trip would prove. Some sixteen years later, Coningham would be the commander of the RAF's Desert Air Force in North Africa, and the route he had pioneered in 1925 would be known as the "Takoradi Route", the principal means of getting US-built aircraft to Egypt and a vital lifeline, not just for the RAF, but for Britain's war effort as a whole.

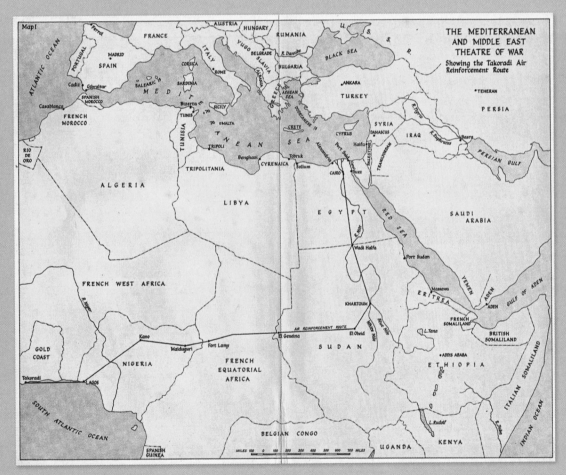

Opposite / An inspection of Avro trainers at the RAF Flying School at Abu Sueir, Egypt, in 1925. The school operated from 1921 to 1929 and the RAF retained a base at Abu Sueir until 1956.

Above / The map shows the Takoradi Route from Egypt to West Africa, which remained a key air supply route for the Allies throughout the Second World War.

Meanwhile, the RAF College at Cranwell had opened on 5 February 1920, while the Apprentice School at RAF Halton had also taken its first intake of students a couple of months later. These mechanics and engineers trained at Halton were known as "Trenchard's Brats" and from the outset became the technical backbone of the RAF, and remain so to this day. Trenchard also established the Auxiliary Air Force – a further contingency to ensure the RAF had enough personnel to operate within the very limited budgets available. The first squadron of "weekend fliers" as they were known, 602 Squadron, was formed in 1925. By then, flying had become a symbol of glamour, excitement and derring-do. The first Hendon Air Pageant had been staged in 1922, and by 1925 had become the Royal Air Force Display, in which the RAF could show off its latest aircraft and also the skills of its pilots. It attracted huge crowds, lots of media attention and helped establish the RAF in the minds of the general public.

By 1925, then, the future of the RAF was secure. Expansion was comparatively slow because of continued limitations on defence budgets, nor did those improve any time soon. In 1928, by which time Churchill was Chancellor of the Exchequer in Stanley Baldwin's Conservative government, the Ten Year Rule was extended, restricting the defence budget yet further. This was followed by the Wall Street Crash of 1929 and the ensuing global depression.

On 1 January 1930, Trenchard stepped down and handed over to Air Chief Marshal Sir John Salmond, although his influence continued. He has been called the "father of the RAF," a term he disliked. He is the man, however, who above all others, ensured the future of the RAF as an independent strategic force. It was because of those strong foundations built in the post-war 1920s that Britain, as the country once again prepared for war in the second half of the 1930s, was able to invest so much of her future war strategy on air power.

Opposite / Flight cadets at the RAF College at Cranwell in 1923. Two of them subsequently became Air Marshals (P. R.G. Bernard, second from left, and G. R. Beamish, far right).

Top / The crew of a Fairey Long-Range Monoplane celebrate a record-breaking flight from RAF Cranwell to Walvis Bay in South-West Africa between February-May 1933.

Above / Fitters refuel an Armstrong Whitworth Siskin fighter of 1 Squadron at Hendon in 1929.

3

THE
BUILD-UP
TO THE
SECOND
WORLD
WAR

Even by the middle of the 1930s, air power was still very new and it remained hard to know exactly what form it would take should large-scale conflict erupt once more. Trenchard had successfully built up the RAF as the third service by making it an efficient means of colonial policing, but he was convinced that a large and effective strategic – that is, independent – bomber force was key to Britain's future air strategy. The damage caused by the limited Gotha Raids in 1917, and by his own Independent Bomber Force, had helped root this view very firmly in his mind. The men that followed him as Chief of the Air Staff in the early 1930s shared it, perhaps unsurprisingly; certainly, the influence of Trenchard continued to be felt long after he formally retired.

Many others also believed the bomber was now the pre-eminent weapon of war. The realization that Britain's island status might no longer be the barrier it had once been made many nervous. In 1932, Stanley Baldwin, then leader of the Conservative Party, famously predicted "the bomber will always get through". Authorities of the age, such as the highly influential Italian General Giulio Douhet, warned that bombers would kill millions in any future war. The Home Office authoritatively claimed that the casualty rate would be 72 people per ton of bombs dropped. This meant that in ten weeks, a bombing force capable of dropping 700 tons a day would lead to 3.5 million casualties.

These were truly cataclysmic figures, and while no-one at the time could prove the truth of such claims, it added a great deal of weight to Trenchard's claim that much of Britain's future military strategy lay with air power. On 19 July 1934, and much to the chagrin of the Army and Royal Navy, who remained broadly ambivalent about the prospect of aerial bombing, the government announced its intention to increase the size of the RAF by some 41 squadrons and for the UK-based Metropolitan Air Force to have 75 squadrons and 1,252 frontline aircraft. Those anxious early days of the RAF, when it had been struggling to survive, let alone have any strategic influence, had, for the foreseeable future at any rate, long gone.

By this time, however, the storm-clouds of war were brewing once more. In 1933, Adolf Hitler and the Nazis had gained power in Germany and a year on it was clear they were already starting to re-arm, in direct contravention of terms of the Treaty of Versailles. Perhaps most troubling were the signs that Germany was rebuilding an air force, which represented another breach of the treaty. Suspicion became fact in March 1935, when Hermann Göring announced the birth of the Luftwaffe. He also boasted that Germany had already achieved air parity with Britain.

The planned increase in size of the RAF announced in July 1934 – Scheme A – was now superseded by Scheme C, which aimed for a further increase of frontline aircraft, and especially in the number of fighters with which to defend Britain. Nor was the government now demanding more biplanes but new, modern monoplanes instead. While speed had not been the only criteria for the RAF's aircraft in the 1920s and early 1930s, it was certainly becoming a considerably more important factor by the time the air expansion schemes were announced.

Interestingly, it was the Americans who were leading the way, and with civilian rather than military aircraft. Air travel over the vast expanse of the USA was clearly very attractive as a means of rapid transportation, and intense competition between civilian airlines soon developed. With it came a demand for high-performance but affordable aircraft. American designers achieved this for the most part by developing greatly improved aerodynamics. This required a considerable initial technical effort, but once this was achieved, it led to simpler aircraft that were then easier to construct and therefore cheaper, too.

Technical teams were sent from Britain to the United States, including, in the spring of 1934, one from Vickers Aviation. Clearly, Britain's aviation industry needed to adopt the American approach to aerodynamics and simpler construction methods, but as Air Marshal Sir Hugh Dowding, the Air Member for Supply and Research on the Air Council, commented, it took "much too long to get an aeroplane through from the design stage to the production stage." Too many committees, too much discussion and too much red tape was stifling the RAF's ability not just to increase in size, but to do so with new modern aircraft.

There was also some debate over the form this increased size should take. A new government under Stanley Baldwin took power in 1935, with Neville Chamberlain as Chancellor. It was he who

Opposite / As war drew near, various plans for RAF expansion were drawn up. Plan A from 1934 (above) called for a frontline strength of 1,544 aircraft; Plan J, two years later (below), for nearly 2,400.

NOTES AND STATISTICS OF R.A.F. PRE-WAR EXPANSION SCHEMES

(Fleet Air Arm figures included up to end of 1937)

SCHEME "A" (for completion by 31. 3. 39.)

DESCRIPTION.

Scheme "A" was designed to provide a maximum first-line strength and, therefore, lacked adequate reserves. Its aim was primarily political in scope and the scheme was meant to "deter" Germany and to impress public opinion at home. No advanced types of Service aircraft were included in the programme.

TABULAR STATEMENT

Date of proposals and authority	Cabinet Approval	M.A.F.		Overseas		Total M.A.F. & Overseas		F.A.A.		Composition of M.A.F. (Bracketed figures indicate Non-regular Sqns.)				Increases in Sqn. I.E.	Provision for Reserves
		Sqns.	a/c	Sqns.	a/c	Sqns.	a/c	Sqns.	a/c	Type	Total Sqns. (I.E)		Total a/c		
16.7.34.	18.7.34.	84	960	27	292	111	1252	16½	213	F	28(5) 12		336	+ 4 a/c	£1,200,000 was to be provided for war reserves up to 1938/9.
C.P.193(34) Interim Report of Ministerial Committee on Disarmament.	Cab (29) 34									LB	25(8) 12		300		Reserves beyond that date were deferred.
										MB	8 12		96	No previous provision	This provision was based on the assump-
										HB	8 10		80		tion that the R.A.F. was not required to
										TB	2 12		24		be ready for war until 1942. The
										GP	4 12		48		Interim Report of Min. Cttee. on Disarmament
										FB	4 4		16		however, pointed out that "The reserve
										AO	5 12		60		must be provided be-
											84(13)		960		fore an outbreak of war becomes imminent".
										Included an Air Striking Force of:- 43 Sqns. of 500 a/c.					C.P.193(34).

SCHEME "J" (for completion by Summer of 1941)

DESCRIPTION

Scheme "J". The Air Staff brief in preparing Scheme "J" was to provide a M.A.F. which would be "(a) a reasonably effective deterrent and (b) enable us to meet Germany as nearly as possible on equal terms". (D.D.(P)12 Memo. of S. of S. for Air, dated 27th October 1937).

The Air Staff plan to achieve "parity" was now conceived in terms of the number and offensive power of the respective bomber types i.e. the yardstick of numerical parity (except as regards the Striking Force) was abandoned.

The M.A.F. bomber force of 1,442 aircraft was meant to achieve parity with the striking force which Germany would possess by the end of 1939. Scheme "J" (Metropolitan) thus envisaged the acceptance of a lag of 18 months in comparison with the corresponding German programme. Only the adoption in peace of a war-system of production, which was excluded by a current Cabinet ruling on the subject, would have avoided this.

Scheme "J" may be regarded as the first expansion programme to be based on calculated estimates of complete strategic requirements[1]. In it our fighter strength was related to "the extent, importance and vulnerability of the areas to be defended" while the Coastal and Army Co-operation squadrons were planned to be adequate to perform their respective tasks of maritime and military co-operation.

In C.P. 316(37), dated 15th December 1937, the Minister for the Co-ordination of Defence, basing his views on financial stringency, accepted only the proposed fighter increases in the M.A.F. He rejected the overseas increases and, while accepting the principle that the striking force might be reasonably increased, suggested that the provision for the reserves should be reduced to allow the arrangements for increasing war potential to be increased. These suggestions were accepted by the Cabinet on 22nd December 1937 and necessitated the production of Scheme "K" which was "J" cut down.

(1) Calculated estimates of the numbers of aircraft required for shipping protection and naval co-operation in case of war with Japan or alternatively for a combined war against Japan and Germany were only made available by the Joint Planning Sub-Committee on 11th October 1937. (C.O.S. 621).

TABULAR STATEMENT.

Date of proposals and authority	Cabinet Approval	M.A.F.		Overseas		Total M.A.F. & Overseas		F.A.A.		Composition of M.A.F. (Bracketed figures indi- cate Non-regular Sqns.)				Increases in Sqn. I.E.	Provision for Reserves
		Sqns.	a/c	Sqns.	a/c	Sqns.	a/c	Sqns.	a/c	Type	Total Sqns. (I.E.)		Total a/c		
12.10.37.	Referred back for modifica-	154 + 4 T.D.	2331 + 56	45	644	203	3031	50 (See D.P. (P)3; for	650	F	38(9) 14		532	+ 3 (+9) + 2	War reserves of a/c for M.A.F. sqns. were to be
D.P.(P)12. Air Staff Memo. "The Requisite Standard of Air Strength".	tion on 22.12.37. Cab. 48							comparison only)		MB	20(7) 21		546	Converted to G.R. @ 21 I.E.	provided on the basis of a/c re- quired (at estimated wastage rates) to maintain
										HB	64 14		896		
										TB	" "		"		

THE WELLINGTON BOMBER

By January 1939, the RAF had 135 frontline squadrons, one more than had been planned back in 1936. All RAF aircraft were now camouflaged (those in Britain with a brown and green scheme). In a further development, squadron numbers and names had been removed and replaced by a two-letter code system painted either side of the fuselage.

RAF Bomber Command had also been equipped with a range of new and modern monoplane bombers, from the Bristol Blenheim, through to the "heavy" Handley Page Hampden and Vickers Wellington. The original specification for a new bomber had been issued back in 1932, and on the back of that the Chief Designer at Vickers, Rex Pierson, along with his deputy, Barnes Wallis, had begun looking at completely new designs. Key to their planned new bomber was a geodetic airframe, which was, in effect, a metal lattice, and which was designed by Wallis and inspired by his earlier work on airships. Covered with fabric, the geodetic design was incredibly strong, but also comparatively light, which meant it could carry more powerful engines and a larger payload of bombs.

The first flight of Vickers' new bomber, designated the Type 271, took place in June 1936. Piloted by Joe Summers, Barnes Wallis was also on board to witness the test flight for himself. With a top speed of 378 km/h (235 m.p.h.), it was armed with three twin Browning machine-guns and capable of carrying just over two tons of bombs. After the success of this first and subsequent further test flights, the Wellington – as it was soon to be named – went into production, with the Air Ministry placing orders for 180 to be built. As with the Spitfire, the Wellington's completely new design caused initial production problems, although by the outbreak of war there were six squadrons equipped with them. It would soon prove an incredibly robust and reliable bomber and in the early years of the war was certainly a match for anything in the Luftwaffe's arsenal. What is more, like the Spitfire, its basic design was so good that it was subsequently further developed and upgraded, and went on to become the only frontline RAF bomber to serve throughout the Second World War.

was now pressing hard for the build-up of defence through fighter squadrons and who argued that the money to fund this should be found through reductions in navy and army budgets. Air Chief Marshal Sir Edward Ellington, the Chief of the Air Staff, thought this was an over-reaction. Like Trenchard, he firmly believed the best way to combat the Luftwaffe was the threat of an even bigger bomber force. Fortunately, Chamberlain, the man controlling the purse-strings, strongly favoured plans to increase the size of the Metropolitan Air Force – as the UK-based RAF was called – and with fighters, too. It was Chamberlain who was in a position to now provide the green light for yet further expansion of the RAF.

The challenges to achieving this were enormous, because suddenly Britain's aviation industry was faced with not only rapidly expanding but also having to use largely new technology as it did so. What was needed was a complete restructuring of the RAF, including the Air Ministry and its procurement procedures, as well as urgently building new and bigger factories and machine-tools with which to construct this more modern force.

It was typical of the top-heavy bureaucracy of the time, however, that Göring's announcement in March 1935 had prompted the creation of yet another committee entitled the "Sub-Committee on Air Parity". The difference, though, was that this committee was headed by the Colonial Secretary, Philip Cunliffe-Lister, who was tough and decisive and immediately injected the kind of clear-sighted drive that was so desperately needed. When Baldwin took over as Prime Minister later that year, he immediately gave Cunliffe-Lister, now elevated to the peerage as Lord Swinton, the job of Secretary of State for Air.

Swinton at once infused the Air Ministry with a new sense of urgency and vigour and was determined not to let progress lag or become mired in lengthy decision making. One of his first

Above / Lord Swinton, the Secretary of State for Air, inspects an RAF recruitment van in 1937. Swinton vigorously championed the expansion of the RAF necessary to fight a new war.

Opposite / Vickers Wellington of 9 Squadron, based at Honington in Norfolk. The squadron flew with Wellingtons from 1939 to 1942.

decisions was to urge government support for the production of two new monoplane fighters designed by Supermarine and Hawker. The Cabinet agreed. Both Supermarine and Hawker had been developing new fighter aircraft in response to earlier procurement specifications. Supermarine, with the success of the Schneider Trophy under their belt, had been favourite to win an earlier fighter contract, but the Type 224 designed by R. J. Mitchell had been a huge disappointment. Supermarine had then gone back to the drawing board. In early 1935, the government had issued another specification for an upgraded fighter plane capable of flying at least 500 km/h (310 mph) and armed with eight machine-guns.

By this time, a new fighter design was emerging from Supermarine. Mitchell himself was already ill with cancer and working on a new four-engine bomber, so the bulk of the design work was left to his colleagues: Beverley Shenstone and Alf Faddy, who developed a new ultra-thin elliptical wing; and Joe Smith, Mitchell's deputy, who did much of the work on how a prototype could be designed for mass production. The Type 300, as it was known, was also harnessed with a new PV12 engine designed by Rolls-Royce, and which provided over 1,000 h.p. Soon renamed the Merlin, the engine was just the power plant Supermarine's new fighter needed.

In the meantime, Sydney Camm, the chief designer at Hawker, was also developing a new monoplane fighter. The tailplane, fuselage and cockpit were based on the latest biplane models,

which made good sense because this meant many of the machine tools needed had already been designed and built, which in turn ensured the aircraft could go into production more quickly. The wings were of stressed metal and Rolls Royce also provided the new Merlin as its power plant. The Hawker Hurricane, as it was named, first flew in November 1935, while the Supermarine Spitfire took to the skies in March 1936. Although each was powered by the Merlin engine, they were very different aircraft, but both offered the speed and armament required. The Hawker fighter would be easier to produce, while the Supermarine Type 300, an entirely brand-new aircraft which, after various revisions, was eventually named the Spitfire, would provide greater production challenges, but had greater speed and rate of climb than the Hurricane model.

The original Air Ministry specification had asked for one new fighter, but both Lord Swinton, and Air Marshal Sir Wilfrid Freeman, the new Air Member for Research and Development, recognized that both planes should go into production. Like

Above, left / R. J. Mitchell, designer of the Supermarine Spitfire, did not live to see his creation help win the Battle of Britain, dying of cancer in 1937.

Above, right / Sydney Camm began his career as a joiner, but rapidly moved into aircraft design, creating the Hawker Hurricane, one of Britain's most successful Second World War aircraft.

Opposite / The prototype of the Hawker Hurricane, which first flew in November 1935. Ultimately, over 14,500 were built, equipping more than half the RAF's fighter squadrons during the Battle of Britain.

Swinton, Freeman had no time for red tape. On 26 May, Humphrey Edwardes-Jones became the first RAF pilot to fly the Spitfire. Nearly crashing on his first landing, he remembered to lower the undercarriage just in time, and emerged from the cockpit unscathed and, as ordered, immediately rang Freeman.

"All I want to know," Freeman said, "is whether you think the young pilot officers and others we are getting in the Air Force will be able to cope with such an advanced aircraft."

"Yes," Edwardes-Jones told him. Within a week, Freeman had placed an order for 310 Spitfires. A contract for 600 Hurricanes had also been made.

These orders became part of repeatedly upgraded expansion schemes, which, by October 1937, had become Scheme J, and which further increased the size of the RAF and called for a total production of some 12,000 aircraft in the following two years, an increase of 170 per cent. Clearly, the Metropolitan Air Force, which included all UK-based squadrons, was already far too unwieldy to cope with this expansion. In 1936, it was decided to replace it with new commands organized by function: Bomber, Fighter, Coastal and Training – and later, in 1938, Balloon Command, which controlled the barrage balloons that floated suspended by wires and prevented enemy aircraft from flying low. Each command had its own staff and headquarters and was divided into geographically located groups. Below the groups were stations and associated airfields at which squadrons were based. General strategy was to be decided by the Cabinet with the advice of the Chiefs of Staff. Responsibility for carrying out that policy rested in the hands of the Air Council and the Chief of the Air Staff.

The number of RAF airfields was also increased to 138 by the summer of 1939, and the land on which to site the new ones was compulsorily purchased. These all required buildings, from accommodation and offices to stores and hangars. New factories and machine tools also had to be built and workers recruited and trained. The challenges were immense, and the story of the initial production of the Spitfire illustrates just some of the many hurdles facing this rapidly growing air force.

Supermarine was a small company based in Southampton and principally made seaplanes and flying boats. It was, however, owned by the much larger Vickers-Armstrong. Despite this, it was initially thought Supermarine themselves would manufacture the initial Spitfire. It became clear, however, that the complexities of mass-production of such a completely new design were beyond the limited size of the Supermarine factory. Entirely new machine-tools had to be designed and then built and only once these were available could the workforce begin training. And only once they were trained could production begin. Consequently, by February 1938, nearly two years after the initial order, not one production Spitfire had been built. The outcry at this

Key to fig.1

Port side of cockpit

1. Flare release controls
2. Signal pistol cartridge stowage
3. Rudder trimming tab control
4. Map stowage box
5. Pressure head heating switch
6. Camera gun master switch
7. Elevator trimming tab control
8. Writing pad container
9. Airscrew pitch control
10. Wedge plate for camera gun footage indicator
11. Radiator flap lever
12. Throttle and mixture friction adjusters
13. Cockpit lamp
14. Mixture lever
15. Push switch for silencing warning horn
16. Throttle lever
17. Boost cut-out control
18. Landing lamp dipping lever
19. Wireless remote controller
20. Main magneto switches
21. Clock
22. Brake triple pressure gauge
23. Landing lamp lowering control
24. Navigation lamps switch
25. Oxygen regulator
26. Flaps control
27. Flaps position indicator
28. Undercarriage position indicator
29. Airspeed indicator
30. Elevator trimming tabs position indicator
31. Reflector gun sight mounting
32. Landing lamps switch
33. Artificial horizon
34. Altimeter
35. Ammeter
36. Rate of climb indicator
37. Direction indicator
38. Turn indicator
39. Gun firing pushbutton
40. Brake lever
41. Control column spade grip
42. Cockpit lamp dimmer switches
43. Compass deviation card
44. Fuel cock lever (top tank)
45. Fuel cock lever (bottom tank)
46. Rudder pedals
47. Rudder pedal leg reach adjusters
48. Seat

F.S./8

Opposite, top / A "D-type" hangar under construction at RAF airfield.

Opposite, bottom / Spitfire construction at Castle Bromwich.

Above / Spitfire manual.

A.P. 1565A, VOL.1, SECT. 1,

PORT SIDE OF COCKPIT

FIG. 1   FIG. 1

F.S./9

revelation was enormous and, sadly, Lord Swinton was forced to resign. However, before his fall, Swinton had developed what had become known as "shadow factories". These were new plants set up to mirror the work of their parent plants in an effort to cope with the kind of mass-production needed for rearmament. In May 1938, Lord Nuffield, the car manufacturer, agreed to build a huge new factory specifically to mass-produce Spitfires along with a new initial order of 1,000. A vast 55-hectare (135-acre) site was chosen at Castle Bromwich near Birmingham and work began immediately on its construction. When completed, Nuffield planned to produce 60 Spitfires a week there.

Cutting red tape, streamlining organization and putting the right people into the right jobs certainly helped transform the RAF in the second half of the 1930s. This was all very much to the good, but production figures were never as high in reality as those laid down in the various schemes proposed by the government. On the eve of war, the biggest constraint on aircraft production was not one of organization, however, but of the technical change from a largely biplane era to one of new, modern monoplanes and stressed metal airframes. Superb new-generation aero engines were being built, but the challenges of retooling, retraining and then producing completely new aircraft could not be mastered overnight.

The early prototype of the Spitfire from 1936. Despite its clear technical prowess, production was slow and by 1938 only one production model had been built.

K 5054

K 5054

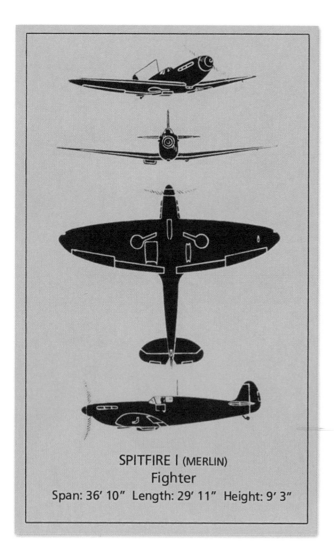

SPITFIRE I (MERLIN)
Fighter
Span: 36' 10" Length: 29' 11" Height: 9' 3"

The first Spitfires finally entered service in August 1938, while work was still underway at Castle Bromwich. By then, many of the problems facing Supermarine had been ironed out and the factory was working at full capacity. By the summer of 1939, 240 of the original order of 310 had been delivered. It was a start, and in these last years of peace, the foundations for much more rapid wartime expansion had been successfully laid. In the years to come, those foundations would not only help save Britain but also help maintain the vital air strategy that would play a critical role in the final victory over both Nazi Germany and Imperial Japan in 1945.

Above / The silhouette of the Supermarine Spitfire became a reassuring presence during the Battle of Britain and it remained in service until the retirement of the last one from the Irish Air Corps in 1961.

Opposite / Introduced in 1932, the De Havilland Tiger Moth became the principal British and Commonwealth trainer aircraft throughout the Second World War.

THE RAF VOLUNTEER RESERVE

The Volunteer Reserve was announced in July 1936 as part of the RAF's new expansion programme and as an extra source of pilots to supplement the Auxiliary squadrons. It was only open to men in civilian life and absolutely no prior experience of flying was necessary. Initially, it was for pilots and needed to take in just 800 entrants a year. Those accepted were paid £25 a year and had to sign on for five years. There were RAFVR centres in all the major cities and instruction, both theoretical and practical, took place over evenings and weekends, including an annual flying course of fifteen days. Recruits had to be aged between eighteen and twenty-five.

One of those who joined the RAFVR was Tom Neil. Tall, with a mop of strawberry-blond hair, he had first become fascinated by aviation during a trip to London with his parents. They had visited Croydon aerodrome and the young Tom had watched the Handley Page 42s with awe. On leaving school in Liverpool in 1937, aged just seventeen, he had tried to join 611 Auxiliary Squadron but had been rebuffed, so on turning eighteen in July 1938, he applied to join the Volunteer Reserve and – much to his great delight – was accepted. "God, life was good!" he noted. "Simply absolutely wonderful."

For the next year, Tom managed to fit his training in around his daytime job as a clerk for the District Bank in Manchester, although there had been little flying training to begin with, as he had joined in October and that winter was a bad one. "England, unhappily," he wrote later, "is not the place in which to learn to fly in winter, particularly if instruction is limited to the occasional weekend." However, by the spring the weather had improved and he was flying in two-seater Tiger Moths. He managed to solo successfully on 20 April, 1939. Later, in May, he underwent his fifteen days' continuous training and then carried on training almost every weekend throughout the summer. On

Sunday, 3 September, the same day Britain declared war on Germany, he presented himself for full-time active service. By that time, he had 60 hours and 15 minutes flying in his logbook, although all of it on Gypsy and Tiger Moths. "And no one had thought fit to give me more than a check circuit and bumps for five months," he wrote. "I couldn't even do a slow roll properly. "Undertrained he might have been, but it had given him and thousands of others a head start now that Britain was at war.

Opposite / Bentley Priory's Filter Room was at the heart of RAF Fighter Command's planning in the Battle of Britain, collating reports of enemy aircraft and passing them to squadrons that needed to scramble.

4

THE DEVELOPMENT OF FIGHTER COMMAND

At the beginning of July 1936, Air Marshal Sir Hugh Dowding was appointed the first Commander-in-Chief of the newly created RAF Fighter Command.

Dowding was Scottish by birth, although had an English father and was educated at Winchester College, after which he had entered the Royal Military Academy at Woolwich and from there was commissioned at the Royal Garrison Artillery in 1900. At that time, Empire policing was very much the prime role of the Army and – like many other young officers – he was posted to the Far East: to India, Ceylon and Hong Kong, although on his return to Britain he decided to qualify for his private pilot's certificate, and then gained his wings with the new Royal Flying Corps. During the First World War, he rose to become commander of 16 Squadron, where he tried to implement new standards of care for his pilots, including regular resting from combat flying. Trenchard disagreed and Dowding was sent back to England; his combat flying career over.

Nor was he initially needed in Trenchard's post-war Royal Air Force, although Dowding's commanding officer at the war's end eventually secured him a permanent commission as a Group Captain and he took on a number of staff roles. By 1930, he was at the Air Ministry as Air Member for Supply and Research and, less than five years later, he was given further responsibilities on the RAF's governing Air Council as Member for Research and Development. At such a time in the development and expansion of the RAF, Dowding's role had been absolutely critical. Not only had he overseen the initial procurement of the Hurricane and Spitfire but also most of those aircraft with which Britain began the war, including what was eventually to become the Short Stirling four-engine heavy bomber. He had also been at the heart of the development of radar.

Dowding appeared to be a rather brusque and stiff individual. He was not such an obviously imposing character like Trenchard, and had a rather cool demeanour, which explains why he was known by the nickname "Stuffy". This was misleading, however. Deeply intelligent, he was also forward-thinking, broad-minded, deeply pragmatic and always prepared to discard convention if he thought it necessary. Not only did he pioneer proper rest for pilots, he was also an early advocate of using radio communication in aircraft and also of issuing pilots with parachutes.

Opposite / HM King George VI and Queen Elizabeth escorted by Air Chief Marshal Sir Hugh Dowding, Air Officer Commanding-in-Chief, Fighter Command, visit Fighter Command HQ at Bentley Priory, near Stanmore, Middlesex, in September 1940.

Below / No. 19 Squadron, based at Duxford, was the first squadron to be equipped with the RAF's newest fighter.

Dowding could certainly be stubborn and was known for his terse and waspish put-downs, but he was also a man who knew his own mind. While there were many in the RAF and Air Ministry who accepted the Trenchard view that the bomber was the pre-eminent instrument of air power, Dowding had seen enough in the last war to know that a bomber force would be slaughtered unless given enough fighter protection. "I have never accepted ideas because they were orthodox," he said, "and consequently I have frequently found myself in opposition to generally accepted views."

Dowding had never been a man to suffer fools, but he was also a profoundly compassionate man, modest, and had a quick wit too. The quiet demeanour also belied an adventurous spirit – one that had led him to be among the first to take up flying but also skiing, then in its infancy, and polo too. Another of his skills was the ability to appoint the right subordinates and to delegate rather than micro-manage. The air defence system he would create worked precisely because it operated on the assumption of delegation and every person in the chain understanding their specific role.

There could be no doubting his qualifications for the job of Commander-in-Chief of the new command, and his most recent job, as head of research and development for the RAF, had made him the obvious person for the post. Yet the task that faced Dowding was enormous, if he was to equip the Royal Air Force to repeal an all-out-attack by their resurgent German counterparts.

Already, however, some of the building blocks for an effective response were in place, not least the development of radar. Nonetheless, training sufficient personnel, pilots and aircrew to use it, while at the same time building the chain of RDF stations and bringing them to a sufficient level of operational efficiency was another race against the clock. Dowding also accepted that radar was not sufficient on its own to secure Britain's air defences. Rather, the radar chain was just one cog. Over the next three years, he recognized it was his task to introduce other cogs and, crucially, to get them working effectively together. His vision, and his ability to harness both technology and the right subordinate commanders, scientists, engineers and civil servants, led him to oversee the first fully co-ordinated air defence system the world had ever known.

By September 1939, after an immense battle against time, this system was largely in place. Another key cog was the Observer Corps. Originally set up in response to the Zeppelin and Gotha Raids during the First World War, it had become a much more refined organization and now a much-expanded one, too. On the eve of war, there were still a few gaps in some of Britain's most extreme reaches, but most of the country was covered by a network of over 1,000 posts and more than 30,000 members of the Observer Corps, volunteers and civilians, although all under direct control of the Air Ministry.

Each part of the country was divided into groups and each group had around 30 posts, which were manned by 14-20 observers provided with equipment to assess the height and numbers of enemy aircraft, and a telephone. Each team worked in shifts and operated in a concentric ring which overlapped with their neighbours so that every part of the sky was covered. Their role was absolutely vital because the information picked up by radar was not especially accurate and in any case ceased the moment raiders crossed the coast. The Observer Corps provided much greater detail about the numbers and height of German aircraft. This information was rung through to the Group Centre, and then forwarded on to the RAF Operations Rooms. The Observer Corps network could handle over a million reports in a 24-hour period, each of which could reach Fighter Command HQ at Bentley Priory in Stanmore, north-west London, in a matter of minutes.

Meanwhile, information from the radar stations was fed directly to the Filter Room at Bentley Priory, where it was logged, analysed and repeatedly updated as plots of enemy raids evolved. The filtered information was then passed on to the Operation Rooms.

Fighter Command was divided into four operational groups. London and the south-east, for example, came under 11 Group, commanded by Air Vice-Marshal Keith Park. Each group was then further divided into Sectors. Each sector had two to three airfields, of which one would be the Sector Station. Headquarters and each group and sector station had its own Operations Room, and every one looked much the same.

At the centre of each was a large map on which both enemy raids and RAF squadrons could be plotted. Overlooking this was a dais on which the controllers sat. On the wall opposite them was the "tote" board, where each sector and squadron was listed with lights that showed their state of readiness or activity. Below were charts monitoring cloud levels and other information. This meant that, at a glance, controllers could be given a very clear picture of what was happening in the air. Those manning the operations rooms were a combination of controllers and plotters, made up from a mixture of RAF officers and WAAFs.

Opposite, top / An Observer Corps unit in action. Their comprehensive coverage of Britain, with local posts in most localities, supplemented radar reports and ensured German planes were rapidly spotted.

Opposite, bottom / The Filter Room at Bentley Priory. Radar reports of attacking German squadrons were plotted on the central table before being passed to RAF Operations Centre for action.

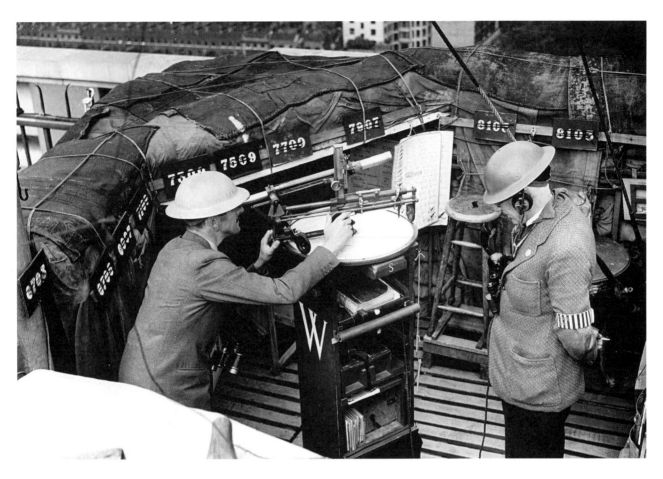

THE DEVELOPMENT
OF RADAR

With many inventions, the science of the day often leads more than one person or team to start developing similar ideas at much the same time. In the 1930s, engineers in both Britain and Germany were beginning to investigate means of creating jet turbines, for example. In much the same way, the principle that high-frequency radio pulses might be reflected off a moving object in such a way that its distance from the origin of the pulses could be measured was being developed in a number of countries in the mid-1930s. These included Germany, but also Britain, where the Scottish scientist, Robert Watson-Watt had begun pioneering a means of displaying this reflecting radio pulse on a cathode ray screen.

Watson-Watt's theories, however, only came to the attention of the Air Ministry by chance. In 1934, Harold Wimperis, Director of Scientific Research at Dowding's then department in the Air Ministry, set up a committee under the well-known physicist Henry Tizard, with the intention that it should investigate the possibilities offered by science to assist air defence. Wimperis initially consulted Watson-Watt about the possibility of

developing gamma rays as a weapon. Instead, Watson-Watt suggested experimenting with what would become known as radar.

Theory was put into practice in 1935 on a cold February day in Northamptonshire, with an RAF aircraft flying in a straight line across a radio beam sent by a BBC short-wave radio transmitter ten kilometres (six miles) away. Sure enough, as Watson-Watt had predicted, there were clear echoes from the bouncing-back of the transmission.

Delighted by the results, Dowding immediately authorized the necessary development money. An experimental station was hastily established at Orford Ness on the Suffolk coast. Within six months, Watson-Watt's radio direction finding, or RDF, as he called it in an effort to dupe the enemy – was detecting aircraft at a distance of as much as 65 kilometres (40 miles). One of the many challenges, however, was how to refine this new-found technology into a practicable form. It was, for example, impossible using only one radio mast to assess the bearing of any incoming aircraft. Only with multiple masts could a picture of the position of a plane at any given moment be achieved. This meant a chain of masts would be needed.

Initially, five radar stations and a training school were ordered to be built. Watson-Watt and his team moved to Bawdsey, south of Orford Ness, in early 1936, by which time he had honed detection techniques further and aircraft were being detected as far as 100 kilometres (62 miles) away. By August 1937, three RDF stations had been built and were functioning and so could be incorporated into Fighter Command's air exercises. There was much trial and error, but the results were broadly encouraging and by then aircraft were being detected

Left / Robert Watson-Watt, who pioneered British radar technology and the means to display its information on a cathode ray screen.

Opposite / Transmitter aerial towers at Bawdsey, Suffolk.

at up to 160 kilometres (100 miles). In the autumn of 1937, the Air Ministry authorized the establishment of a 20-station chain around Britain's coast.

This network, known as Chain Home, however, had its limitations. Watson-Watt had created a system that was entirely static. Each station could only send out and receive signals along the section of coast directly in front of it, and could offer nothing more once an enemy aircraft had passed them. It also required large antennae capable of enough power to achieve the kind of wide floodlight effect that was needed. As a result, CH stations required four 110-metre (360-foot) masts, 55 metres (180 feet) apart, with antenna wires strung between them for transmitting the pulses, and then four different antennae of 75-metre (240-foot) masts for receiving the echo-like reflections. In other words, each radar station was both big and obvious. The British might have been treating radar technology as top secret, but there was no hiding the stations themselves.

It was also possible for aircraft to fly under the beam sent out by Chain Home masts, which meant they would go completely undetected. This, however, was resolved by the addition of a second string of radar stations, known as Chain Home Low, which operated at much lower frequency and sent out on a shorter wavelength, but could be manually rotated by an operator using cranks and which could pick up greater detail, albeit at shorter ranges. The CHL programme was only implemented in the autumn of 1939, and that 30 stations had been built and were operating by June 1940 was largely down to frantic compulsory purchasing of land and a lot of red tape being cut; aircraft production was not the only area where bureaucracy could be overcome when minds were focused.

Radar technology in Britain was, by the outbreak of war, still somewhat rudimentary. Nonetheless, it worked: all around Britain's east and southern coasts, any approaching aircraft could be detected from up to 190 kilometres (120 miles) away. >

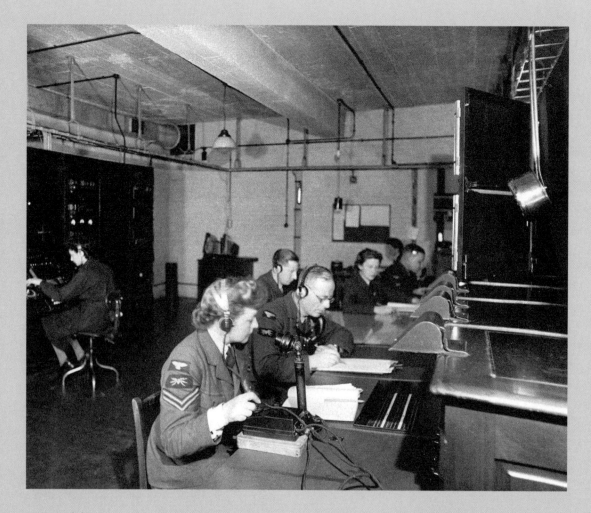

> Meanwhile German scientists had been developing what they called Dezimator-Telegraphie, or DeTe for short. Unlike Watson-Watt's enormous network of masts, the "Würzburg" radar, for example, was highly sophisticated, mechanically rotated and elevated, and capable of guiding both anti-aircraft gunners and fighters onto targets. And since it was rotational, it could operate on a 360-degree setting and over land. They were, however, developed for the Kriegsmarine, the German Navy, and were not used by the Luftwaffe. The Germans also assumed that any other country's radar would be much like their own. This led them to massively under-estimate the effectiveness of the large, ungainly network of masts now strung around Britain's east and south coasts.

Above, top / An east coast Chain Home Radio station in action. On the left an operator plots the course of German aircraft using an RF7 radio receiver.

Above / A Chain Home Radio station at Poling, West Sussex. The station's receiver aerial towers are on the right, the three transmitting aerials are to the left.

The controllers sat on the dais and made the quick decisions about how to best use the information reaching them. In the Filter Room at Bentley Priory, it was the controller's job to ensure the picture compiled was as accurate as possible. This information was then fed to the Operations Room at Bentley and to those at Group level. It was the Group Controllers who would then decide when and how many aircraft should be ordered to intercept an incoming enemy formation. This order would be sent by telephone followed by a teleprinter message to confirm.

Ground controllers were in contact with the pilots through radio. There was other technology used to ensure the controllers could monitor the movements of friendly squadrons, such as High Frequency Direction Finding, or "huff-duff" as it was known (and also "Pip Squeak", the name for the transmissions given off by huff-duff). This meant they could accurately direct, or "vector" squadrons towards the enemy. British planes were also equipped with a homing device called Identification Friend or Foe, or IFF, which gave a distinctive blip if they flew near the coast and were picked up by radar.

Every operations room was duplicated with a second fully functioning room a few miles away that could be used at a moment's notice, while there was also the Defence Teleprinter Network, or DTN, a vast network set up to meet the extra communications requirements of the Air Ministry and also the Admiralty.

Above / Air Vice Marshal Keith Park in a portrait by John Mansbridge, from a series of paintings commissioned by the Air Ministry.

Below / Members of the Observer Corps wait in a sand-bagged emplacement. Once German aircraft had passed the radar stations along the coast, the observers were the principal means to track them.

THE WOMEN'S AUXILIARY AIR FORCE

Although women had served in the Women's Royal Air Force at the birth of the RAF, this had been disbanded in 1920. However, in 1938, with tensions rising and war apparently imminent, the Auxiliary Territorial Service, the women's branch of the army, had been formed. War was averted that autumn, but by the spring of 1939, when Hitler invaded Czechoslovakia, it was clear that Europe lay once more on the brink. With the dramatic increase in size of the RAF and with the development of Dowding's air defence system it was clear there was a vital role for women to play, whether it be in the operations rooms around the country and the Filter Room at Bentley Priory, radar stations or in a host of other staff and intelligence jobs. Not only was it recognized they could do every bit as good a job as men in these roles, by doing so they would be freeing up manpower for combat and more physically demanding tasks.

The Women's Auxiliary Air Force was created on 28 June 1939 with Jane Trefusis Forbes as its head with the rank of Senior Controller, which was later changed to Air Commandant. WAAFs were all volunteers and had to go through initial training before being posted. Among the first to join was Eileen Younghusband, who found herself being posted to the Filter Room at Bentley Priory where she became a plotter. As information from the radar stations was fed by telephone through to the Filter Room, so the controllers filtered that information and the incoming raids – or "plots" – were marked on the large map table by Younghusband and her colleagues using long wooden sticks similar to those used by croupiers in a casino. "The Filter Room was a hive of activity," Eileen

Left / Members of the Air Transport Auxiliary (ATA). Jadwiga Piłsudska (second from right) was the daughter of Poland's pre-war leader Marshal Piłsudski.

Opposite / A recruitment poster for the Women's Auxiliary Air Force (WAAF). Conscription was introduced in early 1941, but over 180,000 of the 250,000 who served in the WAAF were volunteers.

Younghusband recalled. "Girls crowded around the table placing and removing many counters; officers having to push their way through them to put down their arrows or change the information on the metal raid plaques. They looked up to the balcony, answering questions from the officers above. It looked like chaos, but it worked. From the balcony the Controller, a senior RAF officer, was shouting instructions, identifying aircraft, always on the alert."

She greatly enjoyed the experience. The camaraderie was intense and she was glad to feel she was making a difference at the very heart of RAF Fighter Command. Elsewhere, WAAFs played a crucial intelligence role, whether at Air Intelligence at the Air Ministry, at the Government Code and Cypher School at Bletchley Park, for the Y Service, which was a listening station for foreign and enemy radio traffic or as photograph analysts at RAF Medmenham. They were also used as clerks, drivers and nurses.

WAAFs were never employed as aircrew, although WAAF nursing orderlies were flown into Normandy later in the war to help evacuate the wounded. They became affectionately known as the "Flying Nightingales". The only British female pilots in the war were those in the Air Transport Auxiliary – or ATA – which was a ferry service between factories and maintenance depots and home squadrons. They, however, remained civilian pilots rather than part of the RAF.

Conscription for women was introduced in early 1941, and by mid-1943 there were more than 180,000 WAAFs serving the RAF – and by then in more dangerous roles too, as among their number were female agents who were being sent into enemy-occupied territory. Noor Inayat Khan was just one of those who parachuted into France, was captured, tortured and eventually executed in 1944. She was posthumously awarded both the George Cross, the highest award for valour for a non-combatant, as well as the Croix de Guerre.

SERVE IN THE WAAF
WITH THE MEN WHO FLY

Also falling under Dowding's control were the ack-ack guns (anti-aircraft artillery) and searchlight formations, although these were still part of the army. When Dowding had taken over as Commander-in-Chief Fighter Command, there had been only 60 usable but mostly obsolescent ack-ack guns and 120 searchlights. These numbers also needed to be dramatically increased. So, too, did the number of barrage balloons, which floated on wires above urban areas, key factories, ports, railheads and other potential bombing targets. A further handicap was the lack of all-weather metalled runways, a shortcoming which had prevented pilots from carrying out much flying or training during winter months. The Air Ministry had argued that concrete airstrips made airfields more conspicuous from the air. Dowding persisted, however, and eventually got his runways.

The struggle to prepare Britain's home defences had been a constant battle against time. Dowding's doggedness and the not infrequent sharp rejoinders he used in his correspondence with the Air Ministry won him enemies, but by the outbreak of war, his system was working and ensured that if Britain were to come under threat from the air, Fighter Command would be able to provide a fully co-ordinated response. Nowhere else in the world had a system like this and it gave Britain's defenders an enormous – and even a decisive – advantage. It was still not as ready as Dowding would have liked, and nor was it perfect, but it was the only one in the world and collectively, the various cogs in the system added up to considerably more than the sum of their individual parts. Nonetheless, in September 1939, Dowding was well aware that the real test of his system would only come once Britain was attacked.

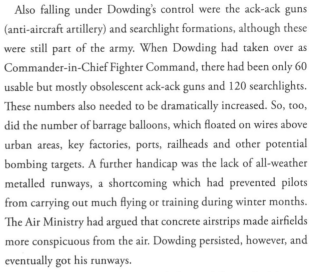

Left / Barrage balloons float high over London, with Buckingham Palace in the middle distance. Low-flying enemy aircraft would become tangled in the balloons' cables, damaging them or bringing them down.

Opposite / A Spitfire loops dramatically in *Victory Above the Clouds*, a painting commissioned by the War Ministry to celebrate the RAF's success in the Battle of Britain.

5

THE SECOND WORLD WAR–BRITAIN UNDER TRIAL

On the very first day Britain entered the war, Sunday 3 September 1939, RAF Bomber Command sent 27 bombers out across the North Sea to search for the German fleet. As they neared Wilhelmshaven, on Germany's narrow North Sea coast, they hit bad weather and then anti-aircraft fire from below showed through the cloud, causing the formation leader to bank and head for home, taking the entire formation with him. All touched back down safely. The first British bomber raid of the war was over. It had achieved not very much.

Among those who had been taking part was 21-year-old Flying Officer Guy Gibson, fresh back from leave on the last day of August. Gibson had been flying a Handley Page Hampden, a twin-engine bomber much like those in the Luftwaffe's arsenal. He had been so nervous before taking off that he had run to the toilet four times and just before take-off had realized his hands had been shaking like a leaf. Once airborne, Gibson felt a little better, his energies focused on flying and keeping formation. Nonetheless, it had seemed surreal to him that he was heading to drop bombs on German targets and that Britain was now actually at war.

The next day, Bomber Command tried again and this time some Bristol Blenheims hit the heavy cruiser *Admiral Scheer* and the cruiser *Emden,* but lost five of their own number. This was to be the last aggressive operation for a while. The French did not want to prompt any retaliation from the Luftwaffe – not yet, at any rate – while as far as Britain was concerned, it made sense to try both to conserve and build its strength for when this would be really needed. For the time being, Bomber Command dropped warning leaflets over Germany and that was it. Needless to say, this also achieved very little.

By this time, Britain was producing around 8,000 aircraft of all types per year, which was pretty much the same as Germany, although the Luftwaffe had been receiving those kind of numbers for rather longer than the RAF. By the time trainers and losses through accidents and general wear and tear were taken into account, this meant the Luftwaffe had around 2,000 frontline aircraft ready for action, (although British intelligence estimated double that), whereas the RAF had 1,460, of which 536 were bombers and 608 fighters. RAF Fighter Command now had 39 squadrons, but it was agreed with the French that the RAF would send a number of bomber and fighter squadrons to France to bolster Allied defences against any German attack; in the military alliance between France and Britain, the Royal Navy was to play the largest role of the British armed forces, then the RAF, while the Army was to send an Expeditionary Force of just ten divisions (the French contribution would be over 100). British air power, then, was to play a significant part.

The trouble was, the RAF in France was there primarily to offer close air support to the forces on the ground, yet official policy was to avoid direct support and for its separate commands to operate independently. Of course, if there was no policy of close air support, there was no reason for there to have been any training in this. Clearly, not enough thought or planning had been put into RAF operations in France, and so those squadrons sent over were rather hurriedly siphoned off into the British Expeditionary Force (BEF) Air Component and the Advanced Air Striking Force (AASF). Both their titles and roles were vague and remained so.

What is more, unlike in Britain, the French Armée de l'Air had created no such air defence system, had very limited radar and had divided its forces into zones of operation that were in no way mutually supporting. Throughout the rest of 1939 and into the first months of 1940, RAF squadrons practised formation flying and put on shows of strength, but without an enemy to combat, there was little to test the existing plans and little thought of challenging some of these potential shortcomings.

Both France and Britain had been content to sit back and continue to build up their military strength while Germany invaded and then overran Poland. The policy was to sit on the defensive, wait for Germany to attack, check that initial assault and then grind down the enemy through greater production and access to global resources. In other words, the plan was much the same as it had been during the last war.

Opposite / A flight of Bristol Blenheims in flight in 1938; they formed the backbone of the Advanced Air Striking Force in France when war broke out the following year.

Above / Air Vice-Marshal Patrick Playfair, who headed Bomber Command's Advanced Air Striking Force in France at the start of the war.

Hurricane pilots of 87 Squadron practise a scramble drill at their base near Lille. Two weeks after the German invasion in May 1940, the squadron was evacuated to Yorkshire.

The Allies had been considering a pre-emptive strike in northern Norway, however, which their superior navies could easily access. Germany was dependent on Swedish iron ore, much of which was transported via the north Norwegian port of Narvik. Neither ally could agree on the form this plan should take, and there was also nervousness that such a move would upset the Swedes, Norwegians and other neutrals, especially the USA. By the time the Allies finally agreed to mine the waters off Narvik, the Germans had also put together a daring plan to invade both Norway and Denmark and, ironically, both parties planned to launch their respective operations on 9 April 1940.

The German attack caught the Allies on the wrong foot, although an RAF Coastal Command Sunderland flying boat had spotted German warships off the Norwegian coast on 8 April. Early the following day, Denmark was swiftly overwhelmed, followed soon after by southern Norway and, although in April the Royal Navy decimated much of the German fleet and in May Narvik fell to British and French troops, in central Norway the British forces arrived too late and found themselves fighting against much shorter German supply lines and against a Luftwaffe able to operate from Norwegian and northern German airfields. Bomber Command

sent squadrons to bomb the German fleet and mine estuaries in Northern Germany, but their results were limited to say the least. Coastal Command also flew repeated reconnaissance missions, but these operations were in no way decisive and no match for the much larger effort by the Luftwaffe. What was urgently needed was air support for the Allied troops heading into central Norway. The problem was getting aircraft there, as it lay way beyond the range of any Hurricane or Spitfire operating from Scotland. The alternative was to fly them to Norway off an aircraft carrier, which involved sending obsolescent Gladiator biplanes of 263 Squadron,

which in turn meant this could never be anything more than a costly symbolic gesture. Although the Gladiators managed to safely land on a frozen lake, most were predictably and swiftly destroyed on the ground by the Luftwaffe. Within a week, just one Gladiator was left and that had been grounded through lack of fuel.

Opposite / Fairey Battle crews check a map in northern France in January 1940. The light bombers would suffer an appalling loss rate at the hands of the Luftwaffe.

Above / Gloster Gladiator biplanes were already obsolete when the war broke out and performed badly in Norway. Most squadrons were rapidly re-equipped with Spitfires.

The RAF hoped for greater success on the Continent, where, by the beginning of May 1940, they now had a total of 38 squadrons operating as part of the Air Component and Advanced Air Striking Force. On paper, the Allies had air parity with the Germans, but the reality was rather different. French squadrons were dispersed throughout the country in different air zones, which operated with a pitiful lack of co-ordination. There was certainly no air defence system as such. When the Germans struck on Friday 10 May 1940, the Dutch and Belgian air forces also now joined the Allies as the northern enemy thrust swept through the Low Countries, but they achieved little; most of their outdated aircraft were destroyed on the ground by the Luftwaffe, who targeted them early and hit them hard before they had a chance to engage.

It was not entirely one-way traffic for the Luftwaffe, however. On 10 May, they lost a staggering 353 aircraft, a single-day toll which would not be exceeded in the next three years. Many of these were transport planes delivering airborne troops, but despite these losses the Luftwaffe quickly gained control of the skies over the battlefield. The German air force had been developed specifically to support ground operations and was formed into air fleets containing bombers, dive-bombers and fighters; there was no separate strategic and independent bomber force such as Bomber Command, for example. Because the French had no air defence system, the Luftwaffe could now concentrate their air power and attack at times of their choosing, sweeping down and bombing and strafing Allied airfields almost at will. In return, the Allies could only hope such airfield attacks took place while their own aircraft

Above / Bristol Blenheim bombers try to blow up the bridges over the Meuse at Maastricht in a desperate attempt to stem the German advance through the Netherlands.

Below / This Spitfire of 222 Squadron was destroyed during a German air raid on Hornchurch in East London, 31 August 1940.

Opposite / A Fairey Battle attacks a German horse-drawn convoy near Dunkirk in the closing stages of the Battle of France.

were already airborne. Allied losses soon mounted; the RAF in France lost 247 aircraft in the first five days, and 857 in May, while the French losses were at almost exactly the same level.

Opportunities were also missed. The bulk of the German armoured forces, which amounted to a mere sixteen divisions out of 135, were the spearhead of the main enemy thrust and – as part of their deception plan – were sent through the wooded and hilly Ardennes region of south-east Belgium, from where they intended to burst across the French border and cross the River Meuse in three places. In getting this mass of vehicles through the narrow roads, which were few in number, gridlock inevitably ensued, a situation which was duly picked up and reported by Allied reconnaissance planes. The French commanders, however, refused to believe what they were being told, as they considered the Ardennes impassable to armour, and so they did nothing. As a result, a golden opportunity to decimate the German spearhead – and with it end the war – was lost.

The RAF did send over 70 aircraft to Sedan on 14 May, but by that time it was too little too late, because the Germans had already managed to cross the River Meuse and establish a bridgehead. They had also brought up more than 300 anti-aircraft guns, which were now assembled in and around Sedan. It was a slaughter, and by early evening wrecked aircraft were strewn all about the wooded slopes of the hills overlooking the town. Of the 71 aircraft sent to Sedan that afternoon, 40 never returned. No single RAF operation of similar size has ever suffered a higher rate of loss in the service's entire history.

Elsewhere, bomber losses were also hideous. Fairey Battles, single-engine light bombers, were dropping like flies. Only one of eight made it back from one bridge-bombing mission on 11 May. The next day, 12 Squadron lost five Battles, 103 Squadron lost two, 105 Squadron another and had two damaged on the ground; 150 Squadron lost two and 218 Squadron another two. The Blenheims were not faring much better. Six were shot down in 15 Squadron alone and three badly damaged. Pilot Officer Arthur Hughes was a Blenheim pilot with 18 Squadron. Three out of four of his fellow crews were killed on 12 May. "I am not really panic stricken," he noted in his diary, "but at intervals a horrid fear seems to seep into my entrails and my stomach grows hollow." By 21 May, when the remains of 18 Squadron were finally posted back to England, just four pilots remained.

In fact, by the time the Meuse front had been blown open, it was clear France was going to lose the battle. The Dutch surrendered on 14 May and all momentum was with the Germans, whose spearheads were able to operate with greater speed than the defenders. The mighty French army, vast though it might have been, was simply unable to bring any kind of co-ordinated

Left / No.1 Squadron at their base in France. All pre-war regulars, they encapsulated the insouciant culture of many fighter pilots; some are even wearing slippers.

Opposite / A Hudson of Coastal Command flies along the French coast at Dunkirk, as burning oil storage units obscure the beaches where the BEF was trapped by the rapid German advance.

counter-attack to bear; lack of radios combined with roads now clogged with refugees had put paid to that, and so the Germans, supported by the Luftwaffe, were able to get in behind them in a giant encirclement, destroying large numbers of French units in detail as they advanced.

The catastrophe unfolding in France prompted much soul-searching amongst Britain's war leaders. Dowding of Fighter Command now urgently pressed to be allowed to bring his squadrons home rather than lose any more in a lost cause. However, while the battle continued, the French continued to urge Britain, their principal ally, for more aircraft, to which the Prime Minister, Winston Churchill, agreed. Nonetheless, events soon overtook these decisions, as German forces took great swathes of territory, including airfields. Often there was no choice but to fly back to England.

By 20 May, German troops had reached the Atlantic coast, and French and British troops were trapped in an increasingly narrow corridor with their backs to the Pas de Calais area of Northern France. Non-essential British troops began to be evacuated and the British drew up plans for the evacuation of as many of the BEF as possible. Air power was to play a vital role in the evacuation and Dowding's squadrons in Fighter Command were to provide that all-important air cover. Fighter Command had been created to defend the United Kingdom, and in many ways the air battle over Dunkirk and Calais in that last week of May and first days of June marks the beginning of the Battle of Britain, as it was the moment when Britain's sovereignty came under threat.

Calais fell on 26 May, the same day Operation DYNAMO, the naval evacuation of the BEF, began from Dunkirk. By this time, the Germans were closing in on the town, with Allied infantry battalions protecting a makeshift perimeter along a system of canals and desperately attempting to hold the enemy at bay. Hitler had given the Luftwaffe a lead role in preventing the evacuation, a task for which they were actually rather ill-suited. Low cloud and smoke from burning oil storage units at the port hid the evacuation beaches and harbour. "The black smoke rose from somewhere in the harbour area," noted Hugh "Cocky" Dundas, a Fighter Command Spitfire pilot with 616 Squadron, "thick, impenetrable, obscuring much of the town." He reckoned the plume rose as much as 4,500

metres (15,000 feet), spreading further the higher it climbed. To make life even harder for the Luftwaffe, most of their units were still operating from airfields far from the coast, which meant they had limited time over the target. What is more, they discovered their Stuka and Junkers Ju 88 dive-bombers, on which they had placed so much emphasis, were not effective when they did not control the air space. As they emerged from their dives, they were both slow and a simple target for Spitfires and Hurricanes waiting to pounce. Nor was it easy for the German bombers to hit comparatively small, moving targets such as ships, which was one of the reasons why so much of the BEF was safely lifted back to England. In all, some 338,000 men were evacuated, and during that time the Luftwaffe lost 326 aircraft compared with Fighter Command's 121.

By the time the evacuation was over, Dowding was extremely concerned about the strength of his command, which had been much depleted by the fighting across the Channel. The worry was that the Luftwaffe would strike Britain straight away, but in fact, Hitler chose to finish off France first. The French, however, did not surrender until 22 June. In that month, Britain built 446 new fighters, which improved Dowding's situation enormously, while in July a further 496 Spitfires and Hurricanes reached his squadrons. The shadow factories were now fully functioning, Spitfire production issues had been ironed out and red tape cut further. In addition, aircraft production had been streamlined to produce just five types:

two fighters and three bombers – Spitfire, Hurricane, Blenheim (also a night fighter as well as day bomber), Hampden and Wellington.

Meanwhile, RAF Bomber Command was now operating over occupied Europe and even Germany itself whenever the weather allowed. The first RAF bombs dropped on German soil had fallen on the night of 15/16 May, when 99 bombers had hit the Ruhr industrial area and a further twelve struck other targets. This and subsequent raids were not especially destructive and certainly did not start the Armageddon so many had predicted before the war. However, they did unsettle the German people who, after the fall of France, had thought the war was all but won. They also rattled the Nazi leadership, who had bragged that Germany was untouchable.

At the same time, the Luftwaffe was hastily building a network of fighter airfields in the Pas de Calais and in Normandy. Their principal fighter, the Messerschmitt 109E, had limited range and so it was important they were based as close to Britain as possible for any planned air assault. This mass of airfields, all bunched together, certainly made a rich target for British bombers, which could hardly fail to miss and which flew over to attack them almost daily.

Opposite / RAF crewmen crowded aboard an evacuation ship during an operation that rescued over 30,000 Allied personnel from Brest on 16–17 June 1940.

Above / RAF fitters repair a Defiant of 264 Squadron in May 1940. The aircrafts' poor performance led to their being relegated to night-time operations by late summer.

◉

Britain had no reason at all to sue for peace. After all, she had the world's largest navy, access to worldwide resources and the biggest empire the world had ever known. Her army might have been defeated, but her navy and air force most certainly were not. There was every reason to feel confident, despite the traumatic and catastrophic defeat of France. Even so, it did not seem quite so straightforward to Britain's war leaders, and least of all to Dowding, whose command could now expect to face the full brunt of the Luftwaffe.

German preparations for an all-out air assault on Britain followed by an invasion, code-named SEALION, were now underway, although Hitler still hoped Britain would sue for terms. Dithering and a lack of inter-service co-operation and focus now marked German plans, and through July, as Hitler prevaricated and Göring moved his air units closer to the Channel, enemy air operations focused mainly on attacking British coastal shipping and trying to draw the RAF out over the water. None of these raids was either large or effective and they went against age-old German principles of concentration of force. In any case, Dowding resisted as far as he could, sensibly preferring to rebuild and conserve his command's strength.

One of the main reasons for Dowding's anxiety was because Air Intelligence had overestimated German strength. British intelligence was both effective and thorough, and they had a clear picture of what units the Luftwaffe had. However, they assumed German squadrons were the same size as those in the RAF. Each RAF fighter squadron, for example, had around two dozen planes and pilots, with no more than twelve airborne at any time. German units had a standard strength of twelve aircraft but often had only nine ready to fly. This meant that, in reality, the Luftwaffe had a numerical advantage against Fighter Command not of 3:1 (the accepted minimum for any attacking force), but more like 1.5:1 compared with the RAF as a whole. However, Dowding's over-estimation of the Luftwaffe's strength was not necessarily a bad thing in a defensive battle.

On the other hand, German intelligence on the RAF was woeful, largely because in a totalitarian state like Nazi Germany, knowledge was power and so the various intelligence bodies were reluctant to

Top / Tracer fire from a Spitfire of 609 Squadron strikes a Heinkel He 111 that was raiding an aircraft factory near Bristol on 25 September 1940.

Right / A German Heinkel's engine blazes after a hit by a 609 Squadron Spitfire near Southampton on 26 September 1940 during the Battle of Britain.

Opposite / A map showing the Battle of Britain Fighter Command area of operations.

FIGHTER COMMAND LAYOUT
JULY 1940

SECRET

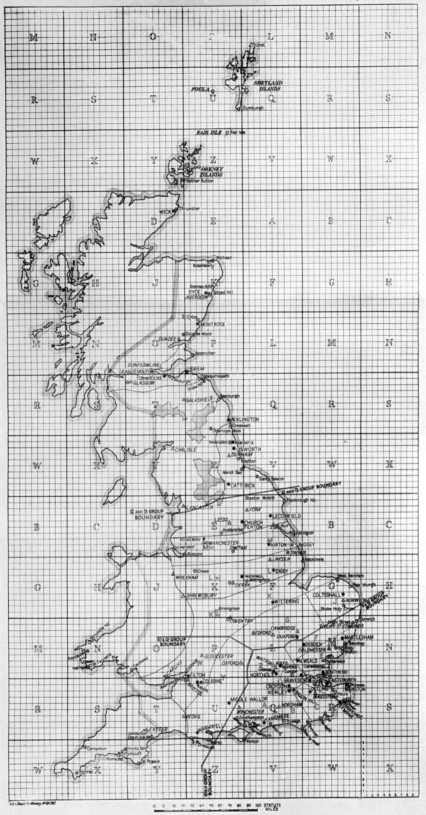

Spitfires from 610 Squadron based at Biggin Hill take flight on 24 July 1940, a day that saw heavy engagements with the Germans over Margate.

share information. Furthermore, all too often, men like Göring were told what they wanted to hear rather than the reality. One of the most senior Luftwaffe intelligence officers, Colonel "Beppo" Schmid, also served as Göring's personal intelligence officer. His qualifications for the post were that he was an early Nazi and a sycophant but not much else; he had travelled little, spoke no English, and had few military qualifications. The intelligence briefings he provided, and on which Göring made his plans for *Adlerangriff* – the "attack of the eagles" – was faulty to say the least. He had no idea the RAF was divided into separate commands, he entirely discounted the ability of the British to repair aircraft efficiently and quickly, grossly underestimated British aircraft production, assured Göring the Hurricane was inferior as a fighter to the twin-engine Messerschmitt 110 (which it was not), and had very little understanding of the radar chain, how it worked or that it was part of a fully co-ordinated air defence system. Yet it was largely based on this intelligence briefing that Göring confidently assured Hitler his Luftwaffe would destroy the RAF given four days of clear weather.

Of course, fighting the RAF over Britain was a very different matter to fighting them in France, something the Luftwaffe greatly underappreciated. Dowding's system had been further honed since the start of the war and almost every enemy raid was now clearly anticipated, plotted and the relevant fighters sent up to intercept, vectored all the while to their targets by ground controllers. The only time Luftwaffe aircraft entered British airspace unseen was if they did so singly or in very small numbers and under the radar. This, however, made them very vulnerable. A lone Junkers Ju 88 twin-engine bomber attacked RAF Middle Wallop on 14 August, for example, and caused considerable damage, but was then swiftly shot down. Accuracy came at a cost.

Göring's all-out assault was officially launched on 13 August and code-named *ADLERTAG* – "EAGLE DAY" – and followed an attempt the day before to destroy the radar chain. Only Ventnor on the Isle of Wight was put out of action, although it was soon repaired. On Eagle Day itself, his bombers reported destroying Eastchurch and a number of Spitfires, but the claim was misleading as it was a Coastal Command station where there were no British fighters. Indifferent weather hindered sustained operations, although there was hard air fighting on both 15 and 18 August. Repeated tactical interference by Göring marked the first two weeks of the battle, which showed no sign of abating within the four days the Luftwaffe chief had hoped. Further

attacks on the radar chain were swiftly abandoned. Fighters were told to close escort the bombers, thus stripping them of their speed, and by 24 August the Stuka force had been so badly mauled it was completely withdrawn.

However, the principal aim for the Luftwaffe remained destroying the RAF on the ground as far as possible, and to that end, airfields were the main target. Unlike in France, however, the RAF was rarely caught on the ground and because most fighter airfields were large and entirely grass, they were comparatively easy to repair; bulldozers, steamrollers and piles of soil were on standby at every airfield. Each airfield also had a secondary control room located a few miles away so if airfield buildings were damaged – and they often were – staff were able to relocate and continue their work without interruption.

Opposite / The ground crew are already refuelling this Spitfire as the pilot climbs down from the cockpit after a sortie in September 1940.

Right / Three teleprinter operators from the WAAF who received the Military Medal for working under heavy Luftwaffe fire during a raid on Biggin Hill on 1 September 1940.

Below / The wreckage of a Spitfire of 222 Squadron lies on the ground after being shot down near Hornchurch in East London on 31 August 1940.

R.A.F. Form 683.

Wt. 43513-4/4141-2   900 M (2 sorts) 3/40.............51/6255

Immediate

SECRET.
CYPHER MESSAGE.

To— AOC in C Bomber Fighter Coastal AOC Flying Training Command AOC 22 group AOC RAF in Ireland RAD BHQ

Date 7/9

Time of 2214 Receipt Despatch

From— Air Min Home forces System

X 322 7/9 ⊙

Serial No. 791

Until 000.1 hours Sept 8 invasion alert no. 2 continues ⊙ Invasion alert no. 1 is introduced at 0001 hrs Sept 8 ⊙ Probable area is SOUTHWOLD to BEACHY HEAD

Received 2250
7 Sept

=1916

CYPHER MESSAGE

ACTION COPY TO

INFORMATION COPY TO

    ,,   ,,   ,,

    ,,   ,,   ,,

DATE

On the night of 24 August, however, Luftwaffe bombers accidentally hit north London and in retaliation, RAF Bomber Command was sent to bomb Berlin the following night. They then hit Berlin a further three times. Once again, they caused little material damage, but it was a reminder that Britain was still very much in the war and that German citizens were not safe. These raids were also a humiliation for Germany's leadership.

By the first week of September, Fighter Command was in a reasonably strong position still, although this has been traditionally seen as the point at which the outcome of the battle became a close-run thing. Plenty of new aircraft were joining the fray, but Dowding and Air Vice-Marshal Keith Park, commander of 11 Group in the south-east, which was facing the brunt of the Luftwaffe's attacks,

were becoming increasingly anxious about the growing loss of pilots. Already, final training had been cut and new pilots were being posted to squadrons without the experience and combat skill they needed. Many of Park's squadrons were now operating at 75 per cent strength.

Above / An internal RAF signal of 7 September 1940 warns of possible German landings on the east coast at a time of very real invasion fears.

Opposite, top / Air Chief Marshal Sholto Douglas inspects men of 303 (Polish) Squadron, the most successful fighter unit during the Battle of Britain, which shot down 126 German planes.

Opposite, bottom / Squadron Leader Brian "Sandy" Lane relaxes with men of his 19 Squadron in September 1940. Lane died after being shot down over the North Sea in December 1942.

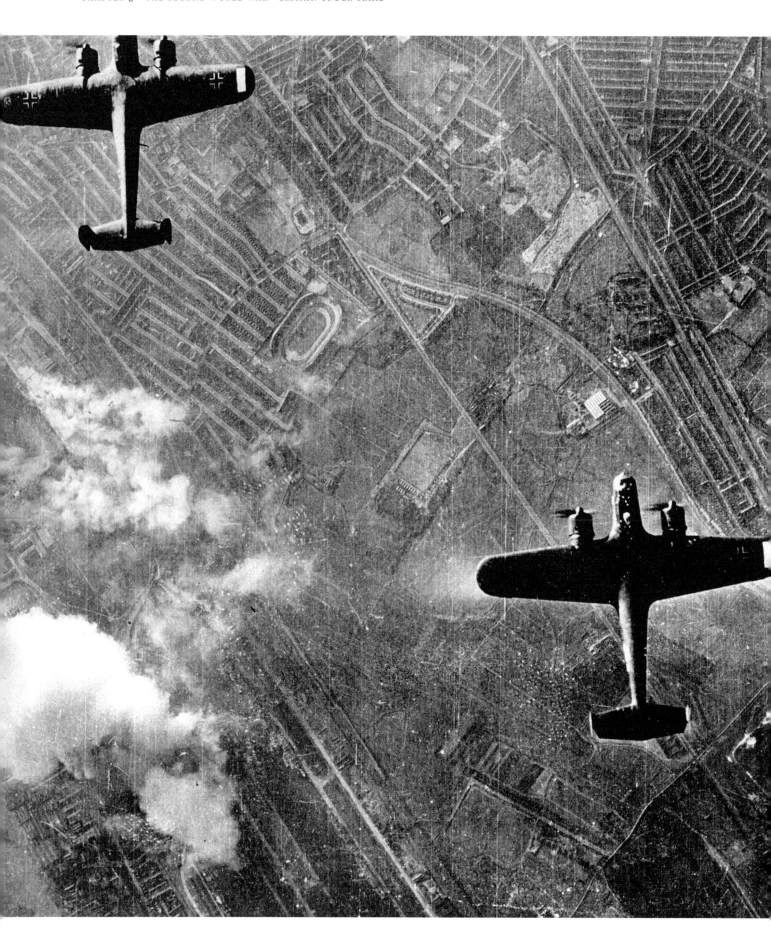

However, even 75 per cent strength still meant 16–18 pilots, and, unbeknown to them, often more than double the number in Luftwaffe squadrons. For the Luftwaffe, the shortage was in aircraft, as production lagged well behind that of Britain and was not keeping up with losses. On 5 September, for example, Siegfried Bethke, a squadron commander in Jagdgeschwader 2 (fighter group), had just three aircraft. "That is my whole squadron," he jotted in his diary. III/JG2, which should have had 36 aircraft, had just twelve. It was a similar picture across most units.

By Saturday 7 September, there were reports of massed barges in the Channel ports and in Britain it was feared that an invasion was imminent. In fact, the Kriegsmarine were still not ready and plans for SEALION continued to evolve rather haphazardly. That same day, Dowding met with Park and other senior commanders to discuss the growing pilot shortage. Park then suggested they implement a system of squadron classification. Those in the frontline with experienced pilots would be classed category "A". Category B squadrons would have a mixture of experienced and newly trained pilots. They would be stationed away from the immediate front line, such as in the south-west. Finally, category C squadrons would have around four experienced pilots and the rest would be new pilots. They would be stationed away from the front line, for instance in the north-east or Scotland. At such a posting, new pilots could build up flying hours, learn from their more experienced colleagues and get some light combat experience against occasional enemy raiders from Norway, for example. Once their hours and training had reached a certain level, they would be posted to a category B or even A squadron.

It was a brilliantly simple plan, saved enormous numbers of lives, and in a trice largely solved the perceived pilot crisis. Also now joining the fray were Polish and Czech squadrons consisting of pilots and aircrew who had escaped to Britain after Nazi Germany had overrun their countries, and who would soon prove very effective. The Polish 303 Squadron, for example, became the highest scoring of the battle. In contrast, the Luftwaffe's declining aircraft numbers could not be so simply and readily solved. This meant that experienced pilots tended to fly more and more with the few aircraft that were left. Dowding, still a firm believer in the importance of carefully managing the fatigue levels of his pilots, insisted each pilot had 24 hours off a week and 48 hours off every three. Few pilots flew more than two sorties – combat flights – a day and very rarely more than four. In contrast, by this stage of the battle, German fighter pilots were regularly flying between five and seven times a day. They were becoming exhausted, both physically and mentally.

Left / Two German Dorniers over West Ham, London, on 7 September 1940, the first night of the London Blitz, which would strike the capital without a break for the next 56 nights.

That same day, 7 September, the Luftwaffe switched tactics and bombed London instead. This has traditionally been seen as saving Fighter Command, but in truth only one airfield out of the 160 in Britain had been knocked out for more than one day and the RAF was still a very, very long way from being defeated. London and other British cities became the main target for the rest of the battle and beyond. The Blitz, as it became known, lasted until May the following year. It did not grind Britain to a halt nor bring about a collapse in morale.

On Sunday 15 September, now commemorated as Battle of Britain Day, there were two sizeable raids on London. The first peaked at around midday and involved some 25 Dornier bombers escorted by 50 fighters. They were attacked by around 280 British fighters. Later that day, a second, larger raid of about 100 bombers and 200 fighters was harried by around 330 Spitfires and Hurricanes. In 11 Group, squadrons were now operating in pairs, but even so, at the point of engagement, most British pilots were one of twelve attacking a much larger formation and this required nerves of steel. Collectively, however, Fighter Command was rarely outnumbered. Furthermore, numbers of pilots and aircraft had actually risen since the start of the battle, not declined.

Operation SEALION was first postponed on 17 September, and then in early October indefinitely so. Enemy raids and fighter sweeps continued, however, but with the threat of invasion over and more and more bombing raids coming only at night, it was clear the RAF had won an absolutely crucial victory. The Luftwaffe had taken a mauling and Germany suffered its first significant defeat. Germany was already running short of resources and Hitler had begun looking to invade the Soviet Union far earlier than he had originally intended. His hope was for a quick victory there, where Germany would benefit from controlling the vast pools of food, oil and other essential war resources, and then turn back to dealing with Britain, and, possibly, the United States. This change of strategy was a major setback to his ambitions, however, and meant he faced a war on two fronts, a situation that had proved fatal to Germany in the last war.

The actual numbers involved in the Battle of Britain were comparatively small, but this should not belie its enormous strategic importance. It was one of the major turning-points in the war, ensured the fight against Nazi Germany would continue, and set Hitler on the long road to defeat. It also more than justified the enormous commitment by the British government to air power that had been begun by Neville Chamberlain and others five years earlier, and set the pattern for future strategy by which air power would be at the heart of Allied plans to defeat the Axis powers.

THE MINISTRY OF AIRCRAFT PRODUCTION

The Ministry of Aircraft Production was created on 17 May 1940, just one week after Neville Chamberlain resigned and Winston Churchill took over as Prime Minister. The man who became the MAP's first minister was Max Aitken, Lord Beaverbrook, a Canadian-born press baron and personal friend of Churchill's. He had no practical experience of aircraft production, but he was a tough and uncompromising businessman, and over the long summer of 1940 proved an astute appointment. It was true that many of the foundations for accelerated aircraft production were already in place, but there was no doubt he also added considerable vigour himself and swiftly cut red tape further. "Committees take the punch out of war," said a notice in big letters in his office. He considered himself a man of action and expected those who worked for him to be so too. The telephone was used more than letters and even telegrams. If there was a bottleneck anywhere, he sent one of his staff to resolve the issue immediately. Anyone who seemed too slow or too complacent was fired, the fate which befell Lord Nuffield, builder of the Castle Bromwich Spitfire plant. Labour regulations were thrown out; factories were expected to work continuously.

He also implemented new systems for the salvage and repair of damaged aircraft run by the Civilian Repair Organisation (CRO). He was able to dramatically improve figures by taking over all Air Ministry storage, insisting on the same work ethic as with new production, and by imposing his will over anyone or anybody that got in his way and threatening to resign if anyone higher up the chain challenged him. The Air Ministry was bypassed entirely. Within six weeks of the MAP's formation, Britain was producing 300 new aircraft a week and more than 250 repaired aircraft were re-entering service. These were decisive numbers. Between the beginning of May and the end of June, aircraft production had risen by

62 per cent, repaired aircraft by 186 per cent and repaired engines by 159 per cent.

Beaverbrook also introduced the Spitfire Fund to Britain. This had begun in Jamaica with local people donating money to fund the building of a Spitfire. Other

Top / A collection of stamps sold to raise money for the Spitfire Fund, an initiative sponsored by Lord Beaverbrook under which localities that raised £5,000 could have a Spitfire named after them.

Above / Max Aitken, son of Minister of Aircraft Production Lord Beaverbrook, was a fighter ace who won the Distinguished Flying Cross for shooting down eight Luftwaffe planes in July 1940.

Opposite, above / A WVS volunteer collects pots and pans destined to be melted down to provide aluminium for the production of Hurricanes and Spitfires.

Opposite / RAF servicemen examine the wreckage of a German Messerschmitt Bf 109 in September 1940. While British production increased, the Luftwaffe struggled to replace its losses.

Dominions caught on to the idea and then Beaverbrook announced that £5,000 would buy one of these precious fighters. Organizations, towns, and the donation by individuals of metal pots, pans and railings all played their part. It was more a public relations coup than anything – Spitfires cost nearer £12,000 – but it helped galvanize the nation by bringing people together to help contribute to the construction of more of the world's most glamorous fighter aircraft.

Beaverbrook moved to the Ministry of Supply in June 1941, by which time Britain was producing ever greater numbers of aircraft, including new heavy four-engine bombers. In all, British factories produced more than 15,000 aircraft in 1940, and by the war's end had output a staggering 132,500, some 50,000 more than Germany. The Ministry of Aircraft Production remained the priority for British manpower until the middle of 1943, underlining the central role given to the RAF in Britain's overall war strategy.

Opposite / An RAF pilot stands next to his Curtiss Kittyhawk in the Libyan desert in 1942. The aircraft is painted with shark-toothed nose art.

6

THE SECOND WORLD WAR – THE GROWTH OF BRITISH AIR POWER

As the summer of 1940 gave way to autumn, the days shortened and the weather worsened, so the amount of daytime enemy raids over Britain diminished. By night, however, German bombers continued to pound Britain's cities using a succession of sophisticated radio navigation systems to try and hit their mark. The Blitz would continue until May the following year and while the level of bombing was certainly out of all proportion to what had been delivered by the Gothas and Zeppelins in the previous war, there was no Armageddon as had been predicted by some in the 1930s. Millions did not die, morale did not collapse and, rather than decline, Britain's industrial output increased dramatically.

Nonetheless, the Luftwaffe were relentlessly targeting Britain's major ports and cities. On 14 November, Coventry, an industrial city in the Midlands, was struck by an unusually large formation of some 515 bombers equipped with the new German *X-Gerät* radio navigation system. Conditions were perfect for bombing, with clear skies and a breeze. First, thirteen Heinkel He 111s dropped incendiaries filled with phosphorous, which gave out a shower of sparks and marked the target. Soon after, the main first wave arrived, dropping high explosive bombs and then more incendiaries. Subsequent waves dropped yet more on to the growing conflagration. As a firestorm took root, the heart of the old city and the cathedral were destroyed. The Luftwaffe had just

demonstrated the best formula for achieving greatest destruction: the right weather conditions and a mixture of high explosive and incendiaries dropped in waves.

RAF Bomber Command had struck back immediately with over 100 bombers hitting the northern port of Hamburg in what was the most successful RAF raid so far in the war. Even so, there was a feeling that now the Battle of Britain had been won, more was needed to defeat the Blitz. One of the ways to achieve this was to find countermeasures to the Luftwaffe's radio navigations systems. In fact, the first, *Knickebein,* had already been foiled, while just over a week before Coventry, a Heinkel He 111 had crashed and the *X-Gerät* set on board had been successfully retrieved and analysed. Over the next few months,

Right / Rubble lies strewn in the devastated ruins of Coventry Cathedral after one of the most damaging Luftwaffe raids of the war on 14 November 1940, which killed over 500 people.

Below / A huge pall of smoke rises from St Katharine Dock by the Tower of London, after a large Luftwaffe raid on 7 September 1940 dropped 50 tons of high explosives, totally wrecking the dock.

the "Battle of the Beams" was slowly but surely won by British scientists, and not least by R.V. Jones, the Assistant Director of Intelligence (Science) at the Air Ministry, who headed a team dedicated to finding countermeasures to German radio navigation equipment. Tragically, an answer to *X-Gerät* had not been found before Coventry, but Jones managed to unravel the German system so completely by the spring of 1941 that the British even anticipated the next system, *Y-Gerät,* and had counter-measures in place before the enemy began using it. Before the Blitz was over, so effective had these counter-measures been, the Luftwaffe had almost entirely given up using such navigation methods.

However, Coventry had focused the minds of Britain's war leaders, who felt not enough was being done to counteract the Blitz. To achieve this not only more anti-aircraft guns, but more effective night fighters were required. Air Chief Marshal Sir Hugh Dowding had been working hard to improve the success rate of night fighters and in fact, British science had already made a giant leap in the right direction with the invention of the cavity magnetron. This allowed for shorter-wave radars, which then permitted the detection of smaller objects from smaller antennas. Instead of being an enormous network of lattices, radar sets could now be drastically reduced in size and made small enough to fit on to a ship or even an aircraft.

By the end of the year, the much-improved Ground Control of Intercept (GCI) radar that was fully rotational and so could track enemy aircraft once they passed the coast, was about to become operational. New on-board Airborne Interceptor (AI) radars were also starting to be used by the crew of Fighter Command's night fighters, of which Guy Gibson, transferred from Bomber Command, was one. These new developments had yet to show their worth, but in time they would dramatically bear fruit. The technological advances made by British scientists working for the Air Ministry have often been overlooked, but the ingenious counter-measures against the enemy, as well as the pioneering work on developing new radars and the cavity magnetron were huge leaps forward.

By this time, Dowding was told to step down. The architect of this decision was Lord Beaverbrook, the Minister for Aircraft Production, and, while it has been regarded as a controversial move ever since, Dowding had avoided retirement twice already and was overdue a rest. It was time for some new blood. Unfortunately, though, the new team that took over – Air Marshal Sholto Douglas as Commander-in-Chief Fighter Command and Air Chief Marshal Sir Trafford Leigh-Mallory at 11 Group – had neither the tactical nous nor the strategic vision of their predecessors.

Above, left / The GCI installation at RAF Sopley in Hampshire, part of a new system that from early 1941 allowed German aircraft to be tracked after they had crossed the British coastline.

Above, right / Air Chief Marshal Sir Trafford Leigh-Mallory, who succeeded Sholto Douglas as head of Fighter Command in November 1942 and commanded air operations for the 1944 Normandy Landings.

The success of the night fighters improved, but as winter gave way to spring 1941, so Fighter Command's day fighters began attempting to take the fight to the enemy. With much of the depleted Luftwaffe now deployed elsewhere, Fighter Command dwarfed the enemy's numbers on the Continent and so flew over in large formations trying to lure the German fighter force into the air. The roles had been reversed, but for many of the same reasons the Luftwaffe had failed over southern England, so Fighter Command achieved little but rising numbers of casualties. By then, the swollen numbers of fighters were needed elsewhere – and the Spitfires especially – but for the time being they remained jealously guarded in Britain.

After the defeat of France and the retreat of the British Expeditionary Force from Dunkirk, Britain was in no position yet to try and get back across the Channel and attack the Nazi-occupied continent. However, her land forces, drawn not just from Britain but also from the dominions of Australia, New Zealand, South Africa and from India, were able to take the attack to the enemy in the Middle East and Mediterranean theatres. Benito Mussolini, the fascist dictator of Italy, had declared war against Britain and France on 10 June 1940, confident that both countries would swiftly surrender and that he could then painlessly absorb British possessions in North Africa.

Britain did not surrender, however. Rather, the Italian Tenth Army marched tentatively into Egypt and halted and was then counter-attacked by the much smaller British Western Desert Force in December 1940 and routed. British forces also invaded Italian-held Abyssinia in East Africa and forced the Italians into retreat.

RAF Middle East, based in Cairo, had around 200 aircraft in June 1940, although as the threat at home receded, 87 Hurricanes, 41 Wellingtons and 85 Blenheims were sent as reinforcements, a number of which were posted to the tiny island of Malta. This had been in British hands since Nelson's day and lay at the very heart of the Mediterranean, roughly equidistant from British-held Gibraltar in the west and Alexandria in the east. It was also 95 kilometres (60 miles) south of Sicily and ideally placed as a base from which to attack any Italian convoys heading to North Africa. In the summer of 1940, Mussolini sent over desultory bombing raids against the undefended island which had little effect. For the first ten days of Malta at war, the entire air defence consisted of a handful of purloined Royal Navy Gloster Gladiator biplanes,

Below / Hurricanes of 261 Squadron parked at Luqa, Malta, in November 1940. The planes played a key role in preventing the Axis powers gaining air dominance and conquering the island.

although the first Hurricanes reached the island on 21 June and more followed soon after.

A continual strain for Air Marshal Sir Arthur Longmore, the Air Officer Commanding RAF Middle East, was that his command was a vast one and the demands wide and varied. At the end of October, Mussolini's forces also invaded Greece, to which the British had guaranteed support. Longmore immediately sent three Blenheim squadrons and two fighter squadrons of Gladiators. There was also an RAF presence in Palestine, Iraq and East Africa and it was the time taken to build up enough air forces to support the Western Desert Force's planned offensive in Egypt that ensured Operation COMPASS was not launched until December 1940. By that time, Longmore had some 220 fighters and bombers to support the ground offensive, compared with more than 400 planes in the Italian Regia Aeronautica. Despite this numerical inferiority, however, the RAF managed to wipe out that advantage in a week. In the same time, they lost just six aircraft and three pilots.

However, the Luftwaffe had by now arrived in the Mediterranean theatre to shore up their wilting Italian ally. Two German divisions under General Erwin Rommel were to be sent to North Africa and, to ensure their safe passage, Fliegerkorps X was posted to Sicily, from where it was expected to attack both the British Mediterranean Fleet and Malta. Although the RAF on Malta had been reinforced

Opposite, above / A Bristol Blenheim lands at a Greek base after a raid on Italian positions in Albania used to launch the invasion of Greece in October 1940.

Opposite, below / Air Officer Commanding Middle East Sir Arthur Longmore inspects an armoured-car company in North Africa ahead of a counter-offensive in the Western Desert in November 1940.

Above / Bristol Blenheims attacking Maleme airfield in June 1941. Their bombs destroyed a number of the precious German Ju52 transports seen here.

Wellingtons of 37 Squadron undergo refuelling and rearming in the Western Desert in 1941.

RAF COASTAL COMMAND

Also playing an increasingly vital role was Coastal Command, who were fully involved in the all-important Battle of the Atlantic. It was here that the war in the West, at any rate, was effectively decided, because only by winning this campaign could Germany hope to stem the flow of supplies to Britain, and ultimately north-west Europe. Fortunately for Britain, Hitler had not laid enough emphasis on strangling Allied sea lanes and so far there had been nothing like enough German surface vessels to compete with the Royal Navy nor U-boats to come close to interrupting the flow of supplies into Britain. In all, just 1.4 per cent of merchant shipping was lost in the Battle of the Atlantic. The key for Britain and the United States, her new ally since December 1941, was to defeat the U-boat menace once and for all. Air power, or Anti-Submarine Warfare (ASW), was key to this, because U-boats could only move at speed on the surface

and tended to submerge only for attack or if spotted. The more they could be forced to submerge, the slower they became and the less chance they then had to attack Allied shipping. In other words, having an air umbrella forced U-boats to evade, reroute and submerge. By the beginning of 1942, that air umbrella covered the western edge of Britain and Ireland and south from Iceland, where the Allies now had a base, forcing the U-boats further into the centre of the Atlantic, which took them much longer to reach.

Further improvements were not long in coming. By the summer of 1942, Coastal Command had a squadron of Very Long Range (VLR) US-built B-24 Liberator four-engine bombers, while a new advanced high-definition 10cm onboard radar was also entering service; the Germans had still not created their own cavity magnetron and had nothing like the radars which the RAF was now

Above / A Sunderland flying boat of 201 Squadron carries out Coastal Command's last operation patrol of the war, escorting a convoy south-west of Ireland in 1945.

Opposite / A Consolidated Liberator I approaches Northern Ireland after an eight-hour delivery flight across the Atlantic in May 1941.

using. The VLRs also had searchlights on board, called Leigh Lights. They could use their onboard radar to search for U-boats then pinpoint them for attack with their lights. This meant the U-boats then had to remain submerged until they reached 12 degrees west, far out in the Atlantic, which was a further drain on time and fuel. In June 1942, twelve U-boats were sunk, half of them by aircraft. Flying for long hours over the Atlantic was unforgiving work and mentally and physically tough on the crews. Coastal Command has rarely been given the credit it deserves, but in this most important of theatres, its crews were playing an absolutely critical role, and continued to do so until the end of the war.

by further Hurricanes, they were no match for the Messerschmitt Bf 109. The problem was that Hurricanes had a slow rate of climb, and could only ever reach around 4,600 metres (15,000 feet) by the time enemy raiders reached the island, when to have the all-important advantage of height they needed to be at more like 7,600 metres (25,000 feet). As a result, the fighter force on Malta was decimated. In just a few weeks in early 1941, a single squadron from the German fighting group Jagdgeschwader 26 (JG26) managed to destroy 42 Hurricanes for no loss at all. Rarely could the superiority of one fighter plane over another have been so starkly demonstrated. Fortunately for the Maltese defenders, the already overstretched Luftwaffe were needed elsewhere: North Africa, the Balkans, Greece and then, in June 1941, for the invasion of the Soviet Union.

Meanwhile, British troops had been withdrawn from North Africa to be sent to help the Greeks, although this was not enough to stop the overwhelming speed and ferocity of the German invasion there in April 1941. The RAF fought valiantly in that campaign, but like the ground forces, there were simply not enough of them to prove decisive. Among the Hurricane pilots sent to Greece was Roald Dahl, later to become an internationally bestselling children's author. Dahl flew twelve combat sorties in just four days and every time found himself outnumbered by at least ten to one, although he did manage to shoot down two enemy aircraft. "We really had the hell of a time in Greece," he wrote to his mother afterwards. "It wasn't much fun taking on half the German Air Force with literally a handful of fighters … Anyway, I don't think anything as bad as that will happen again."

For Germany, moving southwards from the Reich, first into the Balkans and then Greece, it was, of course, far easier to bring aircraft to bear. In contrast, for the British, aircraft had to be crated and moved by ship to the Middle East or flown from aircraft carriers into the mouth of the Mediterranean. However, the inferior numbers were not because of a shortage of aircraft. Between the end of November 1940 and the end of 1941, British factories built

Above / Workers off-loading a crated Hurricane fighter at the dockside of Takoradi port, Gold Coast, May 1942.

Opposite / Air Chief Marshal Sir Arthur Tedder (left), Commander of the Middle East Air Force, and Air Vice Marshal Sir Arthur Coningham (right), Commander of the Desert Air Force.

3,167 Hurricanes and 2,518 Spitfires. In the last quarter of the year, 559 new fighters were built in Britain compared with 320 in the USA and just 221 in Germany. Here then, in Britain, the situation was entirely reversed, with Fighter Command massively outnumbering the Luftwaffe the far side of the Channel. What is more, Fighter Command was bristling not only with Spitfires, but with new improved Mk V variants which were equipped with superchargers that enabled them to dive quicker, an improved Merlin engine and, crucially, 20mm cannons, which were far more destructive than the old .303 Browning machine-guns. The Me 109 was no longer the world's pre-eminent fighter plane.

After the invasion of the Soviet Union, Fighter Command increased the number and scale of large sweeps over France and the Low Countries. Known as "circuses" and "rhubarbs", they flew some 8,000 sorties between June and August 1941 – one sortie being a single aircraft's operational flight – and claimed 322 enemy shot down. In reality, they destroyed just 81. Three hundred aircraft might have justified the huge effort, but 80 did not, and especially not when there was such a need for many, many more fighters in the Mediterranean and Middle East. As the Germans had shown during their blitzkrieg offensives, control of the skies above the battle zone gave an enormous advantage. Without it, as was proved yet again in Greece, for example, there was a limit to what the ground forces could achieve. The RAF of the 1930s had evolved to operate independently of other services in what is termed a strategic role, whereas the Luftwaffe had developed to provide close air support to their ground forces – that is, as a tactical air force. With active theatres in the Mediterranean and Middle East, the RAF now needed both types of air power: strategic and tactical. They also needed to identify the aircraft, tactics, and organization with which to develop tactical air power – and to do so in quick order.

The RAF also needed the right commanders in the right posts. From 1 June 1941, RAF Middle East had a new commander in Air Marshal Arthur Tedder, a tough, forward-thinking leader who, crucially, had experience as head of Research and Development at the Air Ministry. Also now commanding the Desert Air Force, created to provide direct support for the ground operations of the newly created Eighth Army, was Air Vice-Marshal Arthur "Mary" Coningham, who 15 years earlier had pioneered the Takoradi Route from West Africa to Egypt (see p. 44). Coningham was every inch the dynamic fighting commander and brimming with new ideas and energy.

By the start of Eighth Army's CRUSADER offensive, launched against Rommel's Axis forces in November 1941, the RAF in the Middle East was at least able to concentrate efforts with greater focus. Syria had been captured and rebellions in Iran and Iraq quelled, the latter in no small part thanks to Air Vice-Marshal Harry Smart, who had the foresight to create a rudimentary defence force at RAF Habbaniya and to fit bomb racks to his miscellaneous training aircraft. In East Africa, Italian forces were about to finally surrender in November, although the campaign had been all but over since May. This meant that by November, the RAF in North Africa had 554 aircraft ready to fly, while the Luftwaffe had just 121 and the Regia Aeronautica only 192. Tedder had also introduced new improvements to serviceability at the front. Aircraft Replacement Pools were established with seven days' replacements, which were fed from Maintenance Units.

Closer to the front were situated other pools which held two days' replacements. In just four weeks in October, for example, some 232 replacement aircraft were fed into front-line units. Key to success in battle was maintenance of force; RAF Middle East was pioneering new methods to achieve this while the Axis air forces were standing still and not evolving at all.

New methods – doctrine – were evolving too. If the RAF was to directly support ground forces, then co-operation was essential. There was an acceptance that this had largely been lacking in France, for example, back in May and June the previous year. Broadly speaking, Eighth Army wanted greater protection from Stuka dive-bombers and a fighter umbrella directly overhead. This was not possible, however. There were not enough aircraft, nor was it an effective use of air power. Far better was to try and destroy as many Stukas as possible before they were able to attack troops on the ground. A new directive on Army–Air co-operation was issued by Churchill in September based heavily on concepts of close air support outlined by Tedder. In a nutshell,

it insisted that the Army was to specify the targets and tasks needed and the Air Force was to decide when and how to carry out those objectives. Air Support Control (ASC) units were also established. These consisted of a combined Army and RAF team equipped with trucks and radios attached to each corps. Every brigade would have a Forward Air Support Link – an RAF team equipped with a vehicle and radio. Ground units would make requests, which would then be passed to ASCs who then decided how to act. Air support for CRUSADER was consequently far more effective than any land campaign that had come before, materially contributed to its success, and laid down some of the foundations for the future evolution of air–land integration.

One of the problems facing the fighter forces in the Mediterranean and Middle East was the ongoing lack of Spitfires, the RAF's best fighter. Hurricanes were considered largely obsolescent for Fighter Command and so were offloaded to theatres further afield. However, Fighter Command had for some time been receiving more than enough Spitfires and there was no need to keep them zealously within the United Kingdom. This situation was belatedly accepted early in 1942, and on 7 March,

15 Spitfire Mk Vs finally reached the beleaguered island of Malta. By this time, Malta was being hammered by Axis bombers and fighters in a renewed effort to neutralize its defences. Field Marshal Albert Kesselring had been appointed Commander-in-Chief of Axis forces in the South and he had swiftly identified Malta's importance. Some 77 per cent of all Axis shipping to North Africa had been sunk in November 1941, and Malta-based forces had been almost entirely responsible. If Rommel's army in North Africa

Above / Bristol Beauforts based in Malta start an attack run on an Axis tanker in the Ionian Sea. The German failure to reduce the island cost them dear.

Opposite / Groundcrew refuel and rearm a Spitfire at Takali airfield in Malta in April 1942. The arrival of Spitfires on Malta helped tip control of the skies in the Allies' favour.

was to succeed in capturing Egypt and then the Middle East, he would need uninterrupted supply lines. That would not happen so long as Malta was operating as an effective British base.

By April 1942, Malta had become the most-bombed place on earth. During that month, the RAF fighter pilots on the island struggled to fight back. On five separate days they had just one serviceable Spitfire and on 14 April, none at all. Kesselring's blitz ended by May, by which time his Luftwaffe forces based on Sicily were needed in Libya to support Rommel's planned offensive at the end of the month. More Spitfires arrived – 60 reached the island on 9 May and more followed, and despite the extensive damage, the RAF on the island soon regained control of the skies over the island. The Axis failure to completely destroy the island was to prove a catastrophic mistake.

In the North African desert, Rommel's Panzerarmee attacked Eighth Army along the so-called Gazala Line to the west of Tobruk on 26 May. Eighth Army squandered many golden opportunities to defeat the offensive and were routed in turn. First, the Gazala Line collapsed and then Tobruk itself, which had resisted being besieged for eight months the previous year. 21 June 1942 was a black day for the British Army. Eighth Army now fled back to the Alamein Line in Egypt, just 95 kilometres (60 miles) miles west of Alexandria and worryingly close to Cairo and the Suez Canal Zone.

What saved them from annihilation was the Desert Air Force who provided close air support throughout the battle and decisively so during Eighth Army's retreat. Fighters flew round-the-clock bombing and strafing missions, while bombers pounded Axis positions and fighting columns. A truly international force of British, New Zealanders, Rhodesians, Australians, South Africans and Canadians, the Desert Air Force was able to maintain its immense pressure on Rommel's forces by pioneering new methods of "leap-frogging". Squadrons moved forward and back using prepared landing grounds; as a squadron took off, for example, ground crew would head into trucks, hurrying back to the next landing ground, ready to receive the planes as they landed back down again at their new base. This ensured aircraft were able to continue to fly close to the front, which in turn meant they had more time over targets.

PHOTO RECONNAISSANCE

Also playing a crucial role in the Allied war effort were the Photo Reconnaissance (PR) squadrons, now operating across occupied Europe and throughout the Mediterranean. Ironically, it was reconnaissance for which aircraft had first been used militarily, but while air power had broadened, taking photographs, sometimes from high level and at other times from low and more oblique angles, remained vital. It was the photo-reconnaissance work of Malta-based RAF aircraft, for example, that had enabled the Royal Navy's Fleet Air Arm, (brought back to the Navy in 1937), to strike at the Italian Fleet at Taranto with such success back in November 1940. Photo Reconnaissance units continued to throw eyes on enemy defences, build-up of troops and new installations, but also supplemented the decoding efforts of the Government Code and Cypher School at Bletchley Park; although German codes were repeatedly broken, it was imperative the enemy never realized. News of an Axis convoy across the Mediterranean, for example, might be learned by code-breaking, but a Photo Reconnaissance aircraft would then fly over too. Analysis

of this work was carried out in Cairo, and in the United Kingdom at RAF Medmenham, where increasingly well-trained teams, many of which were made up of WAAFs, would examine and interpret what was being gathered.

Above, top / This aerial reconnaissance photo of the German missile launching pads at Peenemünde in north Germany, taken in November 1943, confirmed the existence of the V-1 flying bomb.

Above / A WAAF flight officer photographic interpreter with two Canadian pilots of a photographic reconnaissance squadron, examining newly developed 8" x 7" film at Benson, Oxfordshire, August 1943.

Back in Britain, RAF Bomber Command was slowly growing and evolving, albeit painfully. The RAF had begun the war with strategic bombing central to its war aims, yet so far no bomber force alone had proved anything like as cataclysmic as many pre-war doomsayers had predicted. Bombers could cause a considerable amount of damage, but they had also proved very vulnerable too. Perhaps the biggest failing, however, was their lack of accuracy.

Certainly, so far in the war, Bomber Command did not seem to be delivering on their promise, despite sending wave after wave of bombers over occupied Europe, often at great cost. A special investigation had been launched into the accuracy of bombing and its results were published in the Butt Report of August 1941. This showed that despite what crews were claiming, only one in three bombers managed to get their bombs within eight kilometres (five miles) of the target. It was a devastating blow for Bomber Command and also posed all kinds of uncomfortable ethical questions; most targets were in built-up areas and so inevitably large numbers of civilians were being hit instead. Bomber Command's standing plummeted and many began asking questions as to whether it was a good use of resources to continue to sacrifice bomber crews and aircraft, not to mention the lives of German civilians, if the bombing campaign was achieving so little. After all, British morale had not come close to collapsing during the Blitz and there was no reason to suggest German morale was likely to do so either. Furthermore, Germany was developing an air defence system built on similar principles to that created by Dowding before the war. Now ranged against British bombers were night fighters, radar, observers, ground controllers and ever-increasing numbers of anti-aircraft guns and searchlights.

Despite the Butt Report, and improving enemy air defence, Air Chief Marshal Sir Charles Portal, Chief of the Air Staff since October 1940, strongly believed that strategic bombing remained key to Britain's future war strategy and that it was something with which the British forces should persevere. In this he had an ally in Churchill, who similarly supported the "steel not flesh" policy and believed that improved science, and bigger, better and more numerous bombers could still make a considerable difference.

Changes were afoot. In February 1942, Air Marshal Arthur Harris took over as Commander-in-Chief of Bomber Command. Tough, uncompromising and, like Portal, convinced of the rightness of strategic bombing, his was another sound appointment. Furthermore, he understood that ahead of him lay a period of preparation and building. New navigational aids were being developed and heavy four-engine bombers were also starting to reach the squadrons: the Short Stirling and the Handley Page Halifax were the first to be introduced and there was now a newer, much improved heavy bomber just coming into production. The Avro Lancaster had evolved from the unsuccessful twin-engine Manchester. By adding two more engines and giving the bomber four Rolls-Royce Merlins, Roy Chadwick's bomber was transformed. Not until the American B-29 entered service at the end of the war was there another bomber that could transport a ten-ton payload; this was five times as much as most German bombers could carry. The first three Lancasters reached 44 Squadron on Christmas Eve, but as Harris was well aware, getting to a point where he had hundreds, let alone thousands, of heavy

Below / A Halifax of 76 Squadron, one of a range of new heavy four-engined bombers that came into service in November 1940, allowing larger, longer-range raids to be mounted on Germany.

bombers would take time. The bigger the aircraft, the longer it took to build and those heavies in service would in any case suffer their own levels of attrition. By February 1942, Harris could call on an average of 374 serviceable aircraft daily, of which just 44 were heavies. It was not anything like enough to destroy the Reich.

At the end of May, however, Harris launched the first-ever Thousand Bomber Raid. The target was Cologne, in the heart of Germany's industrial heartland, the Ruhr Valley. Although daily numbers were still only around 400, by scouring Training Command and borrowing 250 aircraft from Coastal Command, he managed to get 1,047 airborne for his dramatic *coup de théatre*. It was a stunt successfully pulled off, however: Cologne had suffered considerable damage and the German High Command had been gratifyingly horrified. It was also a great public relations success that prompted headlines across the free world. Followed up by two more such raids, these operations saw off much of the criticism and sniping about his command and bought Harris some much-needed breathing space with which to develop his force further.

Harris was also competing with many other and increasing demands on the RAF. It rankled with him that the Middle East appeared to be absorbing so many bombers, but every theatre was feeling the pinch and RAF Middle East had, in fact, little more than a quarter of what Harris was claiming. Yet more aircraft were coming – Britain's aircraft factories were producing more than any country except the United States.

In 1942, the RAF Regiment was also formed to provide ground defence for airfields, particularly those near the front, while reinforcements had been sent to the Far East following the Japanese entry into the war, which represented a catastrophe for Britain. This diversion was a drain not only on the RAF at home but also RAF Middle East; the RAF was now operating on a truly global scale, and in 1942, a staggering 23,671 aircraft were built in Britain alone to meet this need. Singapore, Malaya and Burma had all been lost to the lightning sweeps of the Japanese, but the British fully intended to fight back and an offensive into north-west Burma was planned for the end of 1942 once the monsoons passed. For this, air power would be needed. Once again, it was the otherwise obsolescent Hurricanes that were sent, but so too were transports and bombers such as Beauforts and later Beaufighters. It was realized that in the Far East, where infrastructure on the ground was so limited, air supply would have a more important role than perhaps in any other theatre.

Opposite, above / Smoke rises from central Cologne in this aerial reconnaissance photo of the aftermath of the first Thousand Bomber Raid in May 1942.

Opposite, below / Lancasters of 83 squadron take off from Scampton to join the third Thousand Bomber Raid against Bremen during the night of 25/26 June 1942.

Above / Air Marshal Arthur Harris (centre), the architect of Bomber Command's area bombing strategy against German cities, points to a map during the planning of a raid.

A Lancaster of 619 Squadron in February 1944. The aircraft's introduction gave the Allies a heavy bomber with a payload that dwarfed the Luftwaffe's capabilities.

Average net paid circulation
for April exceeded
Daily - - 2,000,000
Sunday - 3,800,000

DAILY  NEWS

Copr. 1942 by News Syndicate Co. Inc. NEW YORK'S PICTURE NEWSPAPER Trade Mark Reg. U. S. Pat. Off.

★★★★ FINAL

Vol. 23. No. 292   New York, Monday, June 1, 1942★   40 Pages   2 Cents IN CITY LIMITS | 3 CENTS Elsewhere

1250 PLANES RAID COLOGNE; CITY IN RUINS

—Story on Page 3

Tea and TNT. (Associated Press Radiofotos FLASHED HERE YESTERDAY from London) Canadian crewmen of a Halifax bomber, one of the more than 1,250 planes that hamburgered Cologne, Germany, enjoy a cup of tea from a U. S.-donated canteen, after returning to Britain. —*Story on page 3; other pictures on page 21*

Back from Cologne

Pilot Officer W. H. Baldwin (left) of Ottawa and Flight Sergt. R. J. Campbell of Pawling, N. Y., have just returned from the attack on Cologne, history's greatest air raid. More than six million pounds of bombs were dropped on Germany's fifth largest city. Nazis retaliated early today with raids on a historic British town.

Back in North Africa, meanwhile, Eighth Army had managed to halt Rommel at El Alamein in July 1942. However, it was rightly felt that a change of command was needed. General Alexander became Commander-in-Chief Middle East, while General Montgomery took over command of Eighth Army. Monty, as he was widely known, immediately chose to site his Tactical HQ alongside that of Coningham's Desert Air Force HQ. This made perfect sense and was something Coningham had been urging for some time. His Desert Air Force, alongside the strategic bombers of RAF Middle East, would continue to play a critical role. This increasing strength, combined with an organizational set-up and tactics that were superior to those of the resource-starved enemy, became startlingly clear in the defeat of Rommel's last offensive at Alam Halfa at the end of August 1942. Relentless air cover and the pounding of supply lines was then backed up by a huge presence over the battlefield itself. Rommel's armour, for example, was stopped in its tracks by heavy carpet-bombing.

Meanwhile, across the Mediterranean, Air Vice-Marshal Keith Park, one of the heroes of the Battle of Britain, had taken over command of the RAF on Malta. Within ten days of his arrival, he had brought a halt to almost all Axis bombing, while the Canadian fighter pilot, George "Screwball" Beurling, was shooting down more enemy aircraft in a shorter period of time than any other Allied ace of the war. Malta's aerial ascendency was proved once and for all when Kesselring ordered a final blitz in October 1942 in an effort to safeguard Axis supplies to North Africa. The Axis air forces were decimated, which meant, of course, they could no longer support Rommel's army across the sea.

The RAF also played a huge part in Eighth Army's great victory in Egypt at the Second Battle of Alamein, which began later that month on 23 October. Throughout the battle, the work of the RAF was relentless. On 27 October, at the height of the battle, 21 Panzer Division had been repeatedly carpet-bombed, abruptly halting the enemy's only counter-attack. "The RAF," noted one Italian soldier, "always wins." Even Montgomery was quick to acknowledge the part played by the RAF. "They are winning the battle for me," he admitted on 2 November. Two days later, it was all over and the remnants of the Panzerarmee were in full flight.

Alamein marked a turning point partly because it was the first time British land forces had decisively defeated German troops on the ground, but also because it marked the moment British and Allied material advantage became overwhelming. From now on, there really would be no more situations like the one Roald Dahl had found himself in during the Greek campaign. Finally, Britain was able to see the pre-war emphasis on air power really take root.

Opposite / An American newspaper front page marks the first Thousand Bomber Raid on Cologne, showing Halifax crewmen enjoying a cup of tea after their epic mission.

Above, left / A Bren-gun team of an RAF Regiment Anti-Aircraft Flight secure a desert landing ground in Egypt in November 1942.

Above, right / Baltimores of the Desert Air Force attack German Panzers during the battle of Alamein.

Opposite / Hawker Typhoon Ib in flight over the Severn Estuary, 1943.

7

THE LONG ROAD TO VICTORY

Just a few days after the end of the Battle of Alamein, a joint Anglo-US invasion force, code named Operation TORCH, landed in Vichy-French north-west Africa. The United States had slowly but surely been gearing up its war production since June 1940, and had entered into technology sharing and war production deals with Britain, which included aircraft production and aero-engine manufacture under licence. Then, in December 1941, the Japanese attack against the US naval base of Pearl Harbor on Hawaii brought America into the war. Following this, the United States and Britain swiftly met for talks in which they reaffirmed earlier pledges to support a Germany-first priority. This meant that while the United States – and Britain for that matter - would take the attack to Japan right away – the priority effort would be to destroy Nazi Germany. The original intention was to build up US forces in Britain for a cross-Channel invasion later in 1942, but it soon became apparent that this would not be possible: the US Armed Forces had been tiny at the start of the war, were still comparatively small in 1942, and were completely untested. In August 1942, a large Allied force of some 6,000 mostly Canadian troops had raided Dieppe, supported heavily by the RAF. It had proved a fiasco. Half the force were killed or captured and a staggering 106 Allied aircraft shot down.

An invasion of north-west Africa would kill a number of birds with the same stone, however. It fulfilled pledges to the Soviet Union to engage German troops on land in 1942, it would give green American troops a chance to gain some much-needed battle experience, and with the Eighth Army driving from the east and the First Army – which included US II Corps – sweeping from the west, there was an opportunity to defeat Axis forces in North Africa once and for all in a giant pincer.

Air power was to play a vital part in this plan, but although the landings, the first large-scale and co-ordinated amphibious operations, were successful and Vichy-controlled French Morocco and Algeria swiftly overrun, Tunisia, another French colonial possession, was a tougher nut to crack. Hitler, always obsessed about the vulnerability of his southern flank, now ordered a massive reinforcement there, which the Germans were able to carry out because of the far shorter distances between Sicily and Southern Italy and northern Tunisia. Fresh divisions, armour, guns and, perhaps most significantly of all, the Luftwaffe, poured

Opposite / Operation TORCH: the RAF commences operations from Maison Blanche, Algeria, following its capture on the morning of 8 November 1942. In the foreground lie kit bags dumped by members of 3201 Servicing Commando Unit.

Above / A Handley Page Halifax BII Series 1 (Special), of 462 Squadron RAAF, based at Gardabia Main, Libya, in flight in early 1943.

into Tunisia. Bad winter weather of heavy rains also hindered the Allied advance.

On 23 January 1943, Tripoli in Libya fell to the Eighth Army, and soon after, Rommel's Panzerarmee, which had been pursued over 1,500 kilometres (1,000 miles) across North Africa, finally crossed the border into southern Tunisia and positioned itself behind the not insignificant defences there. At the same time, British and American war chiefs were meeting in Casablanca in French Morocco to discuss future war aims, and while those mainly dealt with the longer term, the defeat of Axis forces in Tunisia was also at the forefront of discussions. With the campaign somewhat bogged down, but with both First Army in Tunisia and Eighth closing in on the German defences, the structure of the Allied forces was given an overhaul. Alexander became the new Eighteenth Army Group Commander, while unifying the Mediterranean Allied Air Command was an equally important step. Tedder became overall C-in-C, while the US General Carl "Tooey" Spaatz took command of the new Northwest African Air Forces. The biggest part of Spaatz's command, however, was Air Support Tunisia, soon to be renamed North African Tactical Air Force (NATAF), which contained the

Desert Air Force and most of the fighters and medium twin-engine bombers already in theatre. The man given the job of commanding NATAF was Mary Coningham.

Soon after, Coningham outlined his thoughts on tactical air power at a conference organized by Montgomery in Tripoli in February 1943. "The doctrine that we have evolved by trial in war over a period of many months," Coningham told his audience, "could, I think, be stated in its simplest form as follows: the soldier commands the land forces, the airman commands the air forces; both commanders work together and operate their respective forces in accordance with a combined Army–Air plan, the whole operation being directed by the Army Commander."

This was the doctrine for tactical air power and something Tedder and Coningham had been arguing for over the past eighteen months. Coningham's new deputy was an American, Brigadier General Larry Kuter, who was as one with Coningham's vision. Kuter would later be a key architect of the United States's post-war air doctrine, which included many of the principles to which Coningham adhered. Educating ground commanders took time, however, and although in the subsequent Tunisian campaign

Coningham's air forces took on and then systematically defeated the Luftwaffe, it was not without disagreements with those on the ground. Coningham had a vociferous argument, for example, with General George S. Patton, the II Corps commander, over the lack of umbrella fighter cover. Patton was simply betraying his ignorance and, needless to say, Coningham stuck to his guns.

Hitler insisted on reinforcing Tunisia to the end. With more and more ships being sunk, the air bridge became ever-more important, but having neutralized many of the Axis air fields, and with the ring closing in around north-east Tunisia, Coningham's forces were able to intercept many of these transports before they reached the coast. Also helping the Allied effort against Tunisia was the once-besieged island of Malta, now bristling with fighters and bombers. Between the third week of March and the end of April, 519 Axis aircraft were shot down and nearly twice as many again destroyed or damaged on the ground. In the same period, the Allies had lost just 175.

By the beginning of May, the Allies had undisputed control of the air and Coningham's forces once again played their part in the final land battle. In the Medjerda Valley, the Axis defences were crushed in a devastating demonstration of air-land integration. On 13 May, the Axis forces surrendered; the Allies were masters of the North African shores. Air power had been instrumental in this victory.

Opposite / A Bristol Beaufighter operating out of Luqa in Malta in spring 1943. A capable night fighter, the Maltese base allowed it to support the campaigns both in North Africa and Sicily.

Below / Mediterranean Air Commanders: (l-r): Air Vice-Marshal Arthur Coningham, Lt-Gen Carl "Tooey" Spaatz, Air Chief Marshal Sir Arthur Tedder and Brig-Gen Larry Kuter.

At the Casablanca Conference, the Joint Chiefs of Staff had also drafted a new directive to Air Marshal Sir Arthur Harris, the Commander-in-Chief of Bomber Command. "Your prime objective," he was instructed, "will be the progressive destruction and dislocation of the German military, industrial and economic system, and the undermining of the morale of the German people to a point where their capacity for armed resistance is fatally weakened." In effect, he was given *carte blanche* to pulverize German cities. Factories and industry were still to be targeted, but it was now accepted that collateral damage was inevitable and so it had become part of his objective.

A year into the job, Harris was now almost ready to launch his "all-out" strategic air offensive against Germany. Progress had been slow. The arrival of the US Eighth Air Force had meant finding airfields and supplies for the new Allies, which in turn had held back development of Bomber Command. New navigation aids had taken time to come into service but now, in 1942, there was GEE, which operated through setting radio pulses to converge, and a new

device codenamed OBOE. In late January 1943, H2S, effectively the world's first ground-mapping radar, came into service. Harris also now had a Pathfinder Force (PFF). These aircraft, often in the new twin-engine high altitude and extremely fast Mosquitoes, were equipped with OBOE – and now H2S – and would fly ahead of the bomber stream and drop flares onto the target. Initially, Harris had been resistant to the idea of such a force, as he disliked the idea of an elite unit, but the PFF had soon proved its value and they had now become a vital part of bombing operations.

He now also had improved numbers of heavy bombers (particularly Lancasters), and a number of young, tough and dynamic commanders – men like Wing Commander Guy Gibson,

Above / Spitfires cross over Djerba Island in spring 1943 on their way to support British Eighth Army's attack against the Mareth Line, which established an important bridgehead into Tunisia.

Opposite / Winston Churchill and Franklin D. Roosevelt at the Casablanca Conference in January 1943, which was called to co-ordinate strategy against the Axis.

still only 24, but who had been commanding 106 Squadron for nearly a year.

"I was at last," wrote Harris, "able to undertake with real hope of success, the task which had been given to me when I first took over the Command a little more than a year before, the task of destroying the main cities of the Ruhr."

The first night of his offensive was focused on Essen. OBOE-carrying Mosquitoes of the PFF accurately marked their targets and were followed by some 442 bombers, of which 261 were heavies. The ensuing damage was immense: 65 hectares (160 acres) of destruction, 53 buildings of the Krupp armaments works hit, 3,018 houses destroyed and a further 2,166 badly damaged, and

500 people killed. It was an attack that marked a sea-change in the air war, just as Harris had hoped. And it was also just the beginning.

In July 1943, Harris turned his attention to Hamburg, the largest port in Europe and Germany's second city. At least 200 U-boats had already been built there, and, although the Battle of the Atlantic was effectively won by May 1943, Hamburg remained a prime military and psychologically important target. The raids also employed WINDOW for the first time. This consisted of bundles of tin foil dropped from aircraft, which scrambled enemy radar. It had been developed by R. V. Jones and his team and had been ready for use since April, but had not been used for fear the Germans would then use it against the Allies. This had been a poor

decision, though, because during that embargo period, some 2,200 RAF aircraft had been lost, mostly to enemy radar-assisted defences.

WINDOW proved successful over Hamburg, however. Over four separate nights between 24 July and 4 August, RAF Bomber Command, supported during the day by the US Eighth Air Force, hit the city with 4,307 aircraft and some 10,815 tons of bombs. A huge firestorm wrecked the heart of the city, destroying more than 16,000 apartments and killing over 40,000 people, only fractionally less than had been killed in the entire nine-month Blitz. The bombing of Hamburg showed the Nazi regime and people of Germany the devastating destruction that could now be achieved by Allied bombing. Almost four years into the war, the scale of bombing was finally reaching the kind of levels feared in

the inter-war period. Portal and Harris – as well as their American counterparts – had believed strategic bombing alone could win the war. Had Germany been led by a different leader, this might perhaps have proved the case, yet despite the vast amount of damage being caused and the astonishing levels of destruction, Hitler and the Nazi regime vowed to fight on. Allied armies would still be needed to claw their way across Europe, and that meant ever-more dependence on the fast-developing tactical air forces, too.

Above / A De Havilland Mosquito of 544 Squadron in flight. Originally conceived as a bomber, the Mosquito's superior speed led to its adoption as a versatile fighter.

Opposite / A Lancaster on a bombing raid against Hamburg in late January 1943. The operation saw the first use of the H2S airborne ground-scanning radar system.

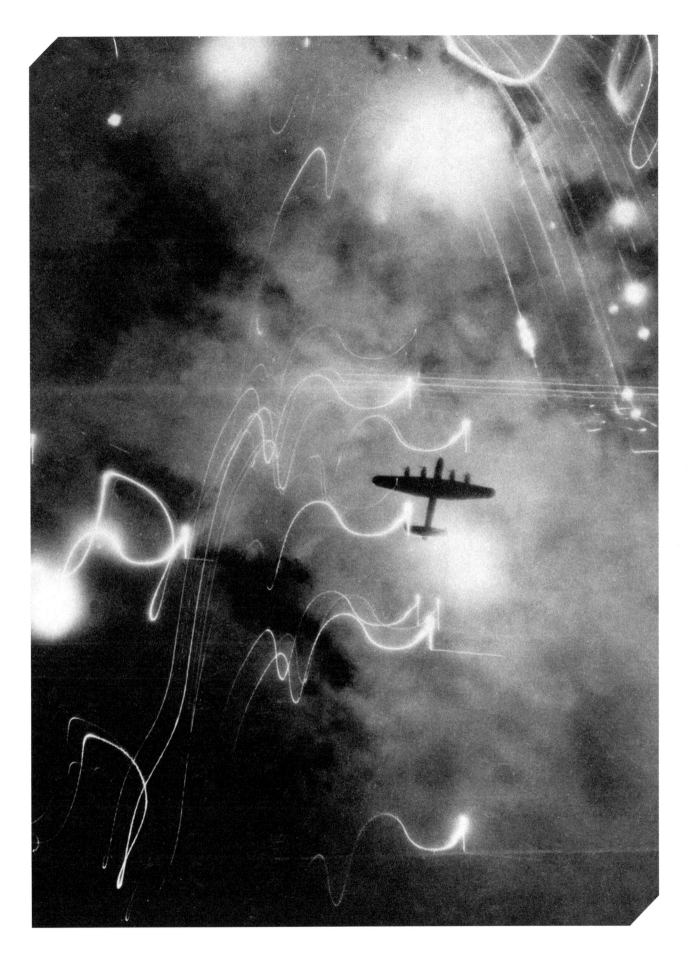

THE DAMS RAID

MÖHNE DAM

Left / An aerial reconnaissance photo of 17 May 1943 shows water pouring through the Möhne Dam after its breach by Barnes Wallis's bouncing bombs the night before.

Below / Wing Commander Guy Gibson, commander of 617 Squadron, whose raid on the Ruhr and Eder dams was one of the most daring aerial operations of the war.

Opposite / A Lancaster of 617 Squadron releases an "Upkeep" bouncing bomb during the raid on the dams on 16–17 May 1943.

The Dams Raid of the night of 16/17 May 1943 is probably the most famous air raid ever launched. Barnes Wallis, the Assistant Chief Designer at Vickers Aviation, had invented a bomb that could bounce on water. This meant it could, in theory, skip over the vertical steel anti-torpedo nets that protected the largest of Germany's gravity dams. These dams, the Möhne especially, but also the Eder, were vital for supplying both drinking water for the large conurbations of the Ruhr industrial heartland and other areas, but also for industrial processes. Their collapse would also cause untold damage. The idea was that the bomb, code-named UPKEEP, would skip over the nets, hit the face of the dam wall, sink and then explode. The effect of the detonation combined with the pressure

of the water above would be enough to destroy these two largest of Germany's dams. However, they had to hit their mark perfectly dead-centre of the dam wall and at a time when the water level above the impact point was high enough to cause the right combination of explosive and pressure. That meant before the beginning of June, and realistically, since they would have to approach the dams at very low level, they would also need the light of a full or nearly full moon. That required a date in the middle of May, or else postponing until the following spring. Wallis had originally devised the bouncing bomb to attack the German surface fleet skulking behind torpedo nets in the Norwegian fjords, and had initially had considerably more interest from the Admiralty than the Air Ministry. By the time Operation CHASTISE, as the raid on the dams was code-named, was given the green light, it was late February, which gave the crews carrying out the raid just ten weeks to get ready.

The man asked to form and lead a special new squadron for this task was Wing Commander Guy Gibson, who had just finished his tour with 106 Squadron, had 72 bombing missions to his name – 22 more than required before being retired from active flight duty – as well as his stint as a night fighter, and was long overdue a complete rest. Gibson hid his mental and physical exhaustion well and accepted the request to lead this one special mission, even though at the time he had no idea what it was.

Gibson arrived at RAF Scampton just north of Lincoln on 21 March, by which time he had barely any crews or any equipment and about nine weeks in which to get ready. Bomber crews usually flew out of close formation and at heights in excess of 5,500 metres (18,000 feet), and then dropped their bombs on what they hoped was the right target. If they detonated within a mile or so they were doing well, despite the Pathfinder Force. Now, he was being asked to oversee the training of an entire squadron to operate at very low level and drop an entirely new type of bomb on a metaphorical sixpence, and in just eight weeks of training. It was a very tall order.

> The squadron was just about ready on time. Gibson had had less chance to practise because he was the only one let in on the secret and he had needed to attend various meetings and trials, which had taken him away from Scampton. Furthermore, he was suffering from gout and the medical officer had recommended he be grounded. To make matters worse, his faithful dog, a constant companion, had been run over and killed the night before the raid.

Incredibly, though, 617 Squadron managed to destroy both the Möhne and Eder dams and damage a third, the Sorpe, so badly that it had to be drained entirely to be repaired. Gibson himself led the way and flew against the Möhne Dam six times: a dummy run, then his live run, and then four more times to try and draw off flak from those that followed. It was actually the fourth UPKEEP, in a Lancaster piloted by Squadron Leader Melvin "Dinghy" Young, that did the fatal damage to the Möhne, but once it was breached, Gibson then led the rest of the first wave of nine specially adapted Lancasters eastwards to the Eder Dam, which was then subsequently breached too by Australian Les Knight and his crew. By any reckoning, Gibson's leadership had been exceptional. He was awarded the Victoria Cross, and rightly so.

However, eight crews from nineteen never made it back and there has been some criticism ever since that it was a good public relations coup and little more, not least because the dams were all repaired by October that same year. This was true, but the works were carried out at enormous cost in terms of time, money and resources and because it was essential to Germany's war effort that they be repaired before the winter rains. In addition, entire villages were swept away, 12 factories completely destroyed and a further 91 damaged. Twenty-five rail and road bridges were destroyed and a further 91 damaged, while several power stations were also destroyed or damaged and shut down. In all, the repair work cost 756 million Reichsmarks, (about £5.9 billion in today's money). Furthermore, not only the repaired dams, but every single other smaller dam in Germany had to be fitted with new anti-aircraft defences. The impact of the Dams Raid should never be belittled.

Below / Guy Gibson's log book for the day of the Dam Busters raid matter-of-factly records "Attack on Möhne and Eder Dams", and deems the operation "Successful".

These would certainly play their part in the Allies' next target. An invasion of Sicily had been agreed at Casablanca and made good sense, because now assembled in the Mediterranean were vast armies, air forces and naval power, while it had also been accepted there could be no cross-Channel invasion until the following year, 1944. If Italy could be forced out of the war, then Germany would be obliged to send more troops to occupy the Balkans, Greece and the Greek islands as well as mainland Italy itself, which in turn would become a drain on their resources and weaken the forces available to defend France.

Operation HUSKY, the Allied invasion of Sicily, was a fraught undertaking, despite momentum now being clearly with the Allies. Planning, which took place while the Tunisian campaign was ongoing, was based on largely speculative assessments of what enemy strength might be, although the amount Germany had thrown at Tunisia suggested this would be considerable; photo reconnaissance work was vital, but could only achieve so much. Moreover, despite

Below / Allied troops land in Sicily during Operation HUSKY, a massive amphibious assault that began on 9/10 July 1943 and depended on Allied air superiority for its success.

deception plans, Sicily was the single realistic next target for the Allies: only over Sicily could Allied fighter cover be provided.

There were four stages to the air plan: to neutralize the enemy air force, destroy enemy communications such as roads, railways and other installations, to isolate the expected battlefield and to provide close air support to the ground forces. In addition, Allied air forces were expected to support naval operations, provide convoy cover, mount airborne landings, protect their own base and rear areas and offer air-sea rescue as well. In other words, their contribution was to be immense, while the opposition air forces, on paper at any rate, were not inconsiderable: intelligence suggested there were some 1,545 enemy aircraft based on Sicily and the toe of Italy. It was imperative that the Allies had control over the skies when the invasion was launched, especially since yet more Luftwaffe units were being sent south to the Mediterranean.

There were a number of airfields on Sicily, spread along its eastern and southern coast and to the west as well. A considerable part of the Allied air forces' work was carried out before the invasion began. Axis airfields were so badly hit by bombers they were only ever able to operate for a few hours at a time. Pantelleria, an Italian island south-west of Sicily, was heavily bombed and the garrison of 4,000 surrendered almost without a fight, while by the time the invasion was launched on 10 July, wrecked aircraft littered the airfields across Sicily. Interestingly, Hitler diverted much of his air force from the Eastern Front the moment the invasion was launched. Between June and September 1943, the Luftwaffe lost 702 aircraft in the Soviet Union, but a staggering 3,504 in the Mediterranean.

Allied close air support over Sicily was criticized as being too infrequent and lacking accuracy. However, the Germans there, fighting largely without such air support, tended to seek the protection of the many towns that dotted the interior. Air support

was called for, but the ensuing destruction served to block the path of the ground forces rather than open it up; Sicily was not like fighting in the open and largely empty deserts of North Africa. Nonetheless, 590 Axis aircraft were destroyed on the ground, 907 were shot down and over 1,000 captured. It was a significant victory on the ground and an enormous one in terms of the air battle. Britain's pre-war strategy had argued that air power would reduce the need for huge armies on the ground and in so doing save lives; it was proving so.

Mussolini had been toppled on 25 July and the Italians had begun to seek terms; the armistice was due to be signed on 8 September 1943. With Operation OVERLORD, the invasion of France, set for May 1944, and even greater Allied forces now in the Mediterranean, it made sense to invade southern Italy, not least because intelligence suggested Hitler would abandon the peninsula below the Pisa–Rimini line to the north of Rome. What clinched the decision, however, was the prospect of swiftly capturing a network of airfields around Foggia in central southern Italy. This was one of the few flat areas of the country and it was agreed the Mediterranean Allied Air Forces could use Foggia to continue tightening the noose around Nazi Germany. From here, the strategic bombing campaign would be able to attack the southern Reich and the Romanian oilfields on which the Germans were so dependent.

This meant that the priority of build-up went to the air forces. This, combined with truly appalling weather and Hitler's change of mind in favour of fighting for every yard, ensured there was to be no swift Italian victory on the ground. However, the Foggia airfields were in Allied hands by the end of October, so the main objective of the campaign had been achieved.

Opposite / RAF ground crews install balloons as a defensive deterrent against Axis air attack shortly after the Operation HUSKY landings in Sicily on 10 July 1943.

Below / Supermarine Spitfire VCs and IXs of 243 Squadron parked ready for take off at Tusciano landing ground, south of Salerno, while providing air cover for the beachhead on 18 September 1943.

Top / Indian troops gather up supplies air-dropped from RAF Dakotas in Burma as another passes overhead dropping additional materiel.

Above / Frank Wilding and Dudley Barnett (in the cockpit), both 136 "Woodpecker" Squadron pilots in Bengal operating over Burma in South East Asia Command.

Meanwhile, in the Far East, the first British attempt to reinvade Burma had failed, while the RAF in their Hurricanes had achieved parity with the Oscar fighter planes of the Japanese Army Air Force, but not a decisive edge. By the autumn of 1943, the British leadership had been culled in much the same way it had in North Africa a year earlier. Admiral Lord Louis Mountbatten had become the first Supreme Allied Commander South-East Asia Command (SEAC), while General Bill Slim had taken command of Fourteenth Army and Air Chief Marshal Sir Richard Peirse the Air Commander-in-Chief South East Asia Command. All three men understood that air power held the key to any possible success on the ground. Spitfires began arriving in the autumn of 1943, and those squadrons taking them were sent on new intensive air combat courses, while the air defence network of radar, observers and ground control was all greatly improved and refined. By early 1944, the RAF was winning control of the skies over the battlefield, an essential precursor to developing advanced air supply to troops on the ground. This was provided by a combined Anglo-US Eastern Air Command that was able to drop and parachute huge amounts of supplies to forward troops operating deep in the jungle. The Japanese way of war was to infiltrate British positions, surround them, then take all their supplies. The British realized that so long as they held firm and denied the enemy those supplies, the Japanese could be systematically defeated. At the Battle of the Admin Box in the Arakan in north-west Burma in February 1944, this proved to be the case. At the same time, medium bombers and Beaufighters were attacking Japanese base areas. Winning the battle in the air was paving the way for victory on the ground.

Over Nazi-occupied Europe, meanwhile, the bomber war continued. Like Fighter Command, Bomber Command's bases were organized into geographical groups, with 5 Group in the centre around Lincolnshire and 6 Group, mostly Canadian squadrons, further north. The Wellington was finally withdrawn from the main bomber force, which was now entirely made up of four-engine heavies and Mosquitoes. Harris's next focus was on Berlin, although it lay beyond the range of Oboe. Bomber Command also struck the German experimental station at Peenemünde hard in August 1943, as well as a number of other targets. Before Foggia was opened up, Bomber Command had also targeted Milan and Turin in Northern Italy.

Every bomber mission was fraught with risk and casualties were mounting; the attacks on Hamburg alone in July 1943 had cost Bomber Command 125 aircraft. On the night of 3/4 November, 589 bombers – of which 344 were Lancasters – attacked Düsseldorf. One crew, piloted by 21-year-old Flight Lieutenant William Reid of 61 Squadron, was attacked by a German night fighter head-on. Hit in the head, shoulders and hands, he managed to keep flying, even though the Lancaster and the cockpit especially were badly damaged. Soon after, they were attacked again, this time by a Focke-Wulf 190, which raked the Lancaster, smashing the starboard tail plane. It peppered Reid's aircraft with bullets, hitting him once more, killing his navigator and fatally wounding his wireless operator. Despite the damage and his own serious injuries, Reid pressed on until he reached the target some fifty minutes later,

Above / Flight Lieutenant William Reid who received a Victoria Cross for his bravery in refusing to abort a bombing raid on Düsseldorf on 3 November 1943, despite his severe wounds.

and duly dropped his bomb load. With his navigator dead, Reid was forced to rely on his own astral navigation to get him and his surviving crew home. With a shattered cockpit, the temperature was freezing, and with the oxygen supply exhausted, he struggled to remain conscious. As he drifted over the English coast once more, he was only semi-conscious, but managed to revive himself enough to safely reach RAF Shipdham in Norfolk, by then being used by the US Eighth Air Force. Although there was mist on the ground and blood from a head wound was obscuring his view, he still managed to land despite a collapsed undercarriage. After hitting the ground, the Lancaster slewed along the runway before finally coming to a halt. Reid survived and for his extraordinarily heroic efforts was awarded the Victoria Cross.

By this time, the RAF and United States Army Air Force (USAAF) were carrying out the Combined Bomber Offensive: relentless round-the-clock bombing, with Bomber Command attacking by night, and Eighth Air Force from United Kingdom and Fifteenth Air Force from Italy bombing by day. Specifically, they were following a US-proposed directive issued on 14 June 1943 to destroy six target systems and 76 specific targets in what was called Operation POINTBLANK. A key objective of POINTBLANK was to destroy the Luftwaffe, which would in turn make future bombing easier and clear the skies before D-Day, the invasion of France planned for May 1944.

Harris had little faith in daylight bombing and two US raids in particular rather underlined why. On 17 August, the USAAF attacked the ball-bearing plants at Regensburg and Schweinfurt and lost 16 per cent of the attacking force. Then on 14 October, of the 291 bombers sent back to Schweinfurt, 198 were shot down or damaged. This was clearly both unacceptable and unsustainable, but by early 1944 yet further technological developments were emerging. The Americans brought in the long-range P-51D Mustang, which gave them fighter escort all the way to Berlin, while further improved radar navigation such as the British GH system was also coming into service. Sheer weight of numbers also made a significant difference, as both Bomber Command and the US Eighth Air Force continued to grow, despite the high rate of attrition. By January, Bomber Command had an average daily number of 818 available heavy bombers, while by May 1944, that figure had exceeded 1,000. The Americans could call on 1,655 daily by May 1944. Back during the Battle of Britain, the Luftwaffe rarely launched raids with more than 100 twin-engine bombers.

Harris remained convinced that weight of bombs alone could defeat Germany, and there was no question that bombing forced the Nazis to disperse their war industry and to dig more and more factories underground, all at a great cost in terms of time, money and resources. Cities were being pulverized, huge numbers of civilians were becoming casualties and Germany's ability to fight effectively was being increasingly impaired as the war continued. For Hitler, however, the world was a black-and-white place: either Germany would triumph and there would be a "thousand year" Reich, or there would be Armageddon. They were heading for the latter course, but Hitler still had no intention of surrendering.

In the third week of February, the Combined Bomber Offensive launched Operation ARGUMENT, a sustained and concentrated attack on the Luftwaffe both on the ground and in the air. Known as "Big Week", it was one of the biggest air battles ever witnessed – of bomber against flak gun, and of fighter against fighter, and in

which tens of thousands of young men were pitted against each other in a brutal clash of arms. Day and night that week, the bombs seemed never to stop falling nor the guns stop booming, while nearly a thousand aircraft were left strewn, wrecked and in pieces all across Germany. Big Week did not destroy the Luftwaffe but critically set them back and reduced the number and quality of trained and experienced German pilots still further. It also succeeded in ensuring the skies over France were largely clear of enemy aircraft for D-Day, now pushed back to June 1944.

Bomber command also played a vital role in the Allies' Transportation Plan, designed to hinder dramatically Germany's ability to reinforce its forces at the front. Annoyed Harris might have been at this perceived diversion, but his attacks so disrupted communications that the vital German panzer divisions, especially, sent to reinforce Normandy, took far too long to reach the front and often arrived disorganized and badly depleted. It meant a swift and co-ordinated counter-attack had been impossible.

In Italy, the tactical air forces were further honing tactics with the introduction of the cab-rank system and "Rover Davids". On the ground, a Royal Artillery forward observation observer and RAF officer equipped with radio would signal targets to squadrons flying overhead. This ensured response time to attack specific targets was much quicker. The Allied air forces in Italy had also been held back by particularly bad winter weather. However, longer, clearer days by

May, when Operation DIADEM, the battle to capture Rome, was launched, meant that they were able to play a much more significant role, both with interdiction against railways and bridges behind enemy lines and with direct close air support. Germans called fighter bombers "jabos" and they made the lives of the enemy a misery. Both at the front and in the rear, there was nowhere for them to hide. Movement by daylight became impossible.

The RAF and Allied air forces made a significant contribution to D-Day on 6 June 1944. Unlike in Tunisia and Sicily, where British airborne troops had been dropped almost entirely by American aircraft and American crews, for Normandy the 6th Airborne Division was dropped by mostly British crews, and the gliders used, although piloted by soldiers, were towed by the

RAF. British airborne operations were successful, with key targets captured and the crucial east flank secured.

Elsewhere, the use of the strategic air forces on D-Day itself was less successful, with much of the bombing over targets along the invasion beaches inaccurate and ineffective in part due to low cloud. Even so, Allied air power reigned supreme,

Left / On 5 June 1944, the day before D-Day, Handley Page Halifax bombers flank Horsa gliders with which they are to deliver airborne troops into Normandy.

with some 11,590 aircraft flying on D-Day itself, of which two-thirds were British. Among them were the Lancasters of 617 Squadron, the Dam Busters, who flew a skilful mission across the Channel further north in an effort to simulate an invasion fleet on an enemy radar screen. However, the Allied Second Tactical Air Force, commanded by Air Marshal "Mary" Coningham, proved invaluable in delaying the arrival of enemy reinforcements and severely restricting movement. One of the German corps commanders, General Erich Marcks, was killed in an air strike, while Field Marshal Rommel, the army group commander, was

also severely wounded in an air attack by Spitfires. As in Italy, enemy movement by day became almost impossible, while the remnants of two German armies suffered a holocaust as they tried to retreat along narrow roads east of Falaise at the end of the campaign.

Bomber Command continued to pound German industry and cities but also carried out many other operations as well. Minelaying was another key role, and by the war's end, they had laid some 47,307 mines off German ports, Norwegian fjords, and U-boat and torpedo-boat bases. They also targeted the V-1 and V-2 sites. These missiles brought a new reign of terror to London, south-east England and newly liberated cities in north-west Europe. Fighter Command also played its part by shooting V-1s down, or "tipping"

them; by flying right alongside, fighter pilots were able to interrupt the air flow over the missiles' wings, which led to them pitching downwards early over sea or open land.

The RAF was also involved in the vast air lift for Operation MARKET GARDEN, the Allied Airborne Army's attempt to capture key bridges ahead of a rapid advance by Second Army. Sadly, planning for such a major operation was hurried and done without full co-operation or consultation with the RAF.

Opposite / Horsa gliders and parachutes litter the field just after the airborne landing by British 6th Airlanding Brigade, which seized bridges over the River Orne and Caen Canal on D-Day.

Above / An aerial photo of Allied landing craft and scores of Canadian troops attempting to leave Juno Beach on D-Day. The very limited German aerial response hugely aided the Allied break-outs.

The operation ended in failure and the loss of large numbers of highly trained and motivated airborne troops. In truth, the scale of the airborne forces, both American and British, grew out of all proportion to its effective air lift capability; airborne troops were volunteers and among the best in both armies. However, they were delivered into battle by among the least good or well-trained pilots and navigators.

Meanwhile, the tactical air forces continued to keep pace with the ground forces, moving up to new airfields as more territory was liberated. Far from being a busted flush, the Luftwaffe kept flying – their pilots might have been greener and less adequately trained, but they still posed a significant threat. New improved Focke-Wulf 190s and the first operational jet, the Messerschmitt Me 262, proved to be formidable foes. In addition, flying low in support of ground troops, shooting up airfields, targeting trains and other ground targets was extremely dangerous. The Germans had some 15,000 anti-aircraft guns of varying calibres within Germany itself and plenty of light flak in their last stretches of occupied territory. Pierre Clostermann was a French pilot flying Hawker Tempests with 274 Squadron. These aircraft could fly well over 640 km/h (400 m.p.h.), and were armed with cannons and capable of carrying rockets and bombs as well. They were superb fighter aircraft, but the work they were carrying out was lethal. On 20 March 1945, five new pilots arrived, but within three days, three of them were dead. "The old hands," noted Clostermann, "worn out by their three sorties a day, were already hard put to it to save their own skins, let alone look after the newcomers." It was like the dark days of the First World War when new pilots were arriving and promptly being slaughtered before they had unpacked their kitbags. "Frightened by their machines, which they flew with great difficulty," wrote Clostermann, "they got themselves massacred by the flak and the Messerschmitt Bf 109s."

Meanwhile, Bomber Command continued to hit targets in Germany. Most of the Reich's cities now looked like battered shells of their former selves, and not least Berlin. The last Bomber Command raid on the capital came on the night of 20/21 April, by which time Soviet forces were about to enter the city. The capital of the Third Reich lay in utter ruins and responsibility for that, above all, lay with RAF Bomber Command. By the war's end, Bomber Command had dropped nearly a million tons of bombs, more than any other nation by quite some margin.

Left / RAF pilots walk past their Hawker Typhoon fighter-bombers after a successful raid against a German V-1 launch site in France in January 1944.

On 8 May, the war in Europe was finally over, but the fight against Imperial Japan went on. Here, air power continued to play a critical and integrated role. In defeating the threat to India and then in carrying out the subsequent reconquest of Burma, Allied commanders were as one in recognizing its importance. During the epic battle of Imphal from March to July 1944, General Slim first lured the Japanese Fifteenth Army towards the town and its six surrounding airfields in a fighting retreat designed to degrade his attackers, then fought back and eventually largely destroyed the enemy forces. While Japanese supply lines grew ever longer, so Slim's shortened, and although the town was completely cut off by road, supply lines were kept open by air. In the last week of March, 5th Indian Division became the first force of its size to be transported into battle entirely by air, while elsewhere, troops at the front were kept in food and ammunition through localized air drops. Protecting the transports were the indefatigable fighters and fighter bombers. In the spring of 1945, Slim began his last great battle in Burma itself with a two-fisted attack south towards the key city of Mandalay, and a second thrust wide through the jungle to the south

towards Meiktila. This drive by IV Corps was achieved in secret and was possible only because of the RAF's control of the skies and the resupply of IV Corps by Allied air forces the entire way. This was the last and decisive battle in Burma and from then on until the Japanese surrender in August, the enemy were beaten and on the run.

The war's end, after nearly six long years, came on 15 August 1945. Air power had not won the Allies victory in its own right, but had played a vital role. The part played by the RAF had made victory possible, saved countless Allied lives and unquestionably shortened the overall length of the conflict. It could rightly feel proud of its enormous contribution to victory.

Opposite, above / Paratroopers descend on the Netherlands during the first phase of Operation MARKET GARDEN in September 1944. The failure of ground troops to link up with them led to disaster.

Opposite, below / Crews arm Typhoons of 247 Squadron with rockets in Normandy in June 1944. The squadron provided air support as the Allies advanced into Germany.

Below / RAF and WAAF personnel at North Killingholme wave off 550 Squadron Lancaster "F-Fox" at the start of her 100th – and last – mission, a raid on Bochum, Germany, in November 1944.

THE BOMBING
OF DRESDEN

In the last months of 1944 and early 1945, RAF Bomber Command, together with the American strategic air forces, effectively destroyed what remained of Germany's oil industry, pummelling synthetic fuel plants to devastating effect and reducing production to just 10 per cent of capacity. They also hammered the Reichsbahn, the German railways system, which was very much the glue that bound Germany's war effort together. Together this brought Nazi Germany's fighting capacity largely to a standstill and played a vital part in hastening the end of the war.

Moral question marks have remained over the strategic air campaign against Nazi Germany, and despite the massive damage caused to cities such as Hamburg, Essen and Berlin, it is the attack on Dresden on 14 February 1945 that has caused the most soul-searching ever since. It is, however, no good judging what happened during the Second World War through the prism of 21st-century century sensibilities. All sides could only fight the war with the technology and weaponry at their disposal; there were no laser-guided missiles, nor even atomic bombs before Germany's surrender. Any rational and humane leader would have realized years earlier that his country could not win and so saved the lives of his countrymen by accepting an armistice. Instead, the RAF continued to bomb Germany as a means of preserving British and Allied lives and shortening the war. Once Germany surrendered, however, the bombing ended.

The myth has it that Dresden was full of harmless, arts-loving innocents concerned only with manufacturing luxury goods such as cameras and china. In fact, Dresden was a Nazi stronghold before Hitler even took power; Martin Mutschmann became the regime's longest-serving Gauleiter (governor), and certainly one of its most brutal, while anti-Semitism was not only a popular policy, but pursued with gusto. The populace sacked

the synagogue in 1938, one of the city's most beautiful buildings. By 1945, Dresden was home to no fewer than 127 factories employed in war work – work that occupied the vast majority of the population, and which made it a legitimate military target.

By February 1945, Dresden had also become a major railhead, with hundreds of thousands of troops heading to the rapidly approaching Russian front as well as south into Italy. It was for this reason that Dresden was bombed: the British were helping the Russians. That night, as had been the case with Coventry more than four years earlier, the conditions were perfect: the anti-aircraft defences had largely been removed, the sky was clear, and the targeting was as precise as the technology of the day allowed.

In the end, 25,000 people were killed, rather than several hundred thousand as initially claimed. Many died because they stayed down below in shelters, as they had been told to do. Unfortunately, the majority of these shelters were grossly inadequate: as the blaze above absorbed all the oxygen, those huddling together were slowly asphyxiated or poisoned by carbon monoxide.

The bombing continued. Pforzheim was attacked ten days after Dresden, and one in four were killed in the firestorm there, rather than the one in 20 at Dresden. Würzburg was also 90 per cent destroyed by the Allies in February 1945. These decisions were made not just by Harris, but with the support of his superiors, the government and the Americans, and they did shorten the war. This does not make their destructive power less appalling, but does ask questions about accusations of senselessness. It was Joseph Goebbels, the Nazi propaganda chief, who picked on Dresden as his final, most cynical PR scoop. Amazingly, it is the story he spun that most people still believe today.

Opposite / The central goods yard and marshalling depots along the Elbe are still ablaze the day after the massive RAF raid on Dresden on 14 November 1944 caused a fireball that killed 25,000 Germans.

8

THE COLD WAR: 1945–57

The first task of the post-war RAF was to bring home hundreds of thousands of servicemen and former prisoners-of-war as well as help with the immediate aftermath of such colossal destruction and chaos. Displaced people had to be fed and tended, and so food and medical supplies were flown to Europe, the Middle East and the Far East, too.

So much had changed over the last few months of the war and no one in the RAF could have doubted the post-war world would be a very different place. One of the biggest sea changes was technological. During the Second World War, both Britain and Germany had been leading the world in terms of new jet technology. While the Luftwaffe had been first to get a jet aircraft operational, however, this was largely due to an act of desperation. Not only did jet technology offer a chance for superior aircraft, but paradoxically, although jet technology was complicated, actually manufacturing turbojets was simpler than building piston engines. This meant that jet engines could be built using slave labour forced to work deep in an underground factory at Mittelwerk, in Thuringia. Here thousands of Jumo jet engines were manufactured – many more, in fact, than could possibly be fitted into the number of air frames being produced.

In Britain, meanwhile, companies such as Bristol, Rolls-Royce and De Havilland could afford to take longer, not only over research, but also in development, safe in the knowledge the Allies were already winning the war and managing without any urgent need to hurry up jet engine production.

The turbojet had been invented by Frank Whittle back in the 1930s, although it was not until the eve of war in 1939 that he finally

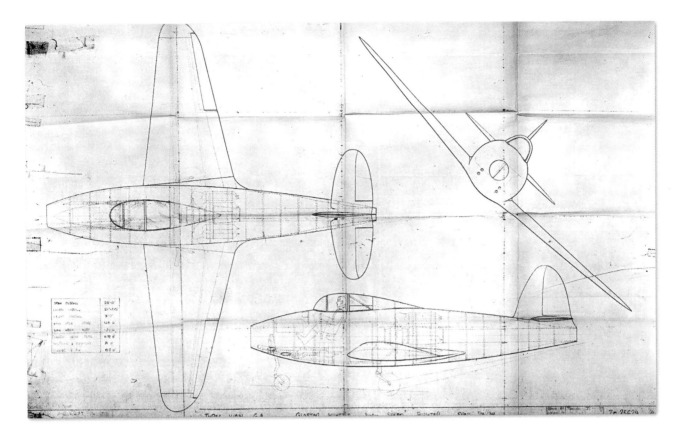

received the backing of the government. The Gloster E28/39 was the first jet-powered aircraft specification issued by the Air Ministry, and the subsequent airplane finally took to the skies on 15 May 1941. Whittle was an undoubted genius, but a less effective businessman; his company, Power Jets, had been created through the backing of a number of venture capitalists and the company was overly dependent on Whittle himself. The W2B turbojet, which powered the first prototype jet, proved unreliable and by this time both De Havilland and Rolls-Royce were not only producing their own engines, but ones that were better than the Whittle W2B. Whittle produced a

new engine, the W2B/500, but it was not taken up and never went into production. By 1944, Power Jets had been sidelined and Whittle was suffering a nervous breakdown.

Opposite, above / Frank Whittle patented a turbo jet engine in 1930, but the British government's slowness in taking it up allowed the Germans to be the first to fly a military jet, in 1944.

Opposite, below / The Gloster E28/39, the British prototype jet aircraft, which first flew in May 1941.

Above / Preliminary drawings for the E28/39. Its successful first flight showed that jet propulsion was viable, but the first operational fighters only reached squadrons in July 1944.

Meanwhile, Rolls-Royce were developing jet engines, and taking them to a new level. Just as important as Whittle in the story of the jet engine is Dr. A. A. Griffith, now almost entirely forgotten, but a brilliant engineer who worked at RAE Farnborough. In 1926, he had produced a seminal paper, "Aerodynamic Theory of Turbine Design". This outlined the design of the axial compressor, later developed further at Farnborough by Griffith's assistant, Hayne Constant. Whittle's jet had been a centrifugal-flow design, which meant the front of the engine was large. As Griffith pointed out, it would be impossible to produce really high-speed engines with such a design. The axial flow used a

mass of angled blades rotating and narrowing to compress gas or air. It was much more fuel efficient and narrower, causing less drag and as a consequence, greater speed. During the war years, Rolls-Royce were able to merge both jet and axial-flow technology and create the first axial-flow turbojet. The Rolls-Royce Avon, for example, was designed in 1944, but first ran in 1947 and powered the new English Electric Canberra from 1950 onwards. The development of jet engines was not a priority for the Allies, as they were not viewed as necessary for winning the war; that was being achieved with the aircraft they already had. In fact, the Gloster Meteor first flew in RAF colours on 27 July 1944, but they were not necessary for combat. Rather, Britain's wartime jet aircraft production was aimed more towards the post-war market.

Certainly, by the war's end, Britain had not only an enormous aircraft industry, but was also the world leader in terms of jet engine technology. The Soviet Union had not developed jet technology at all, and although the Americans had been given a Whittle W2B, they had produced no jet aircraft themselves. Germany, however, had been ahead in terms of airframe design. As one of the war's victors, Britain had sent pioneering test pilots like Captain Eric "Winkle" Brown to speak to leading German aircraft designers and also to test fly many of their new designs. The investigations of men like Brown and others were then applied to new jet technology. Suddenly, Britain had a golden opportunity to dominate the world in terms of future air power and aircraft production. Jet technology promised power, prestige and money: Britain could defend itself with new, vastly superior aircraft, become world leaders in civil aviation too, and rebuild crippled finances by selling this technology abroad. Suddenly, the race was on amongst Britain's numerous aircraft manufacturers to produce new and fantastic aircraft – machines that would become the envy of the world.

It was clear that for all the speed and agility of the Gloster Meteor, for example, the traditional theory of placing an engine in each wing – as had also been the case with the Messerschmitt Me 262 – was not ideal, because jet engines did not behave in the same way as propeller-driven engines. In a jet aircraft, if one engine failed, the other would cause the aircraft to slew around in what was known as "asymmetrical flying". Meteors were also easy to stall during training, especially, with far too many young airmen getting themselves killed.

British firms now began experimenting with new Luftwaffe-inspired airframe designs, some with delta wings, others with swept-back wings, and yet others with twin fuselages. On the ground, British car manufacturing was only very slowly developing, but in the air, Britain's skies were alive with screaming jet aircraft of fantastical futuristic designs hurtling over in a flash of silver. Needless to say, for the test pilots, this period of intense development was exceptionally dangerous. Between 1945 and 1952, some 80 test pilots, most of whom had survived combat in the war, were killed testing new aircraft.

Among those to die was Geoffrey de Havilland Junior, son of the pioneering aircraft designer. In September 1946, de Havilland Jr had been testing the new DH108 jet with swept-back wings and an engine capable of 3,000 lbs of thrust. Chasing the dream of being the first man to break the sound barrier, he climbed high and dived. He had reached Mach 0.875 (1,072 km/h, 666 m.p.h.) when the aircraft broke up. The remains of the DH108 crashed into the mud at Gravesend and de Havilland's body was found a couple of days later. He had died of a broken neck and so had been unable to bail out. Eric Brown was asked to carry out the crash test investigation and re-enacted the flight. At 1, 200 metres (4,000 feet) and flying at Mach 0.88 (1,078 km/h, 670 m.p.h.), his DH108 aircraft suddenly began to oscillate, shaking violently. Another aircraft had been nearby and the pilot had seen Brown's DH108 turn into nothing more than a blur. Fortunately, Brown, who was much shorter than de Havilland had been, was able to regain control despite the immense G-forces. "I realised," he said later, "it had been a very narrow shave." The DH108 had simply not been designed to travel at such speeds. The next generation, however, the DH110, would be.

Meanwhile, the RAF needed to significantly reduce its size. Not only was the war over, but a new era had emerged in which single atomic bombs dropped by one aircraft could achieve the same

destruction as thousands of heavy bombers. The RAF had ended the war with over a million men and women, but by 1947 had been reduced to 300,000. Large numbers of aircraft had been sold or broken up. Sadly, no-one considered it worthwhile to box up samples of each aircraft for posterity and so many of the iconic aircraft of the war were lost forever. Others, however, still had plenty of service left. New improved marks of Spitfire would be operational into the 1950s, as would the Hawker Tempest, for example, while Avro's renowned Lancaster bomber was developed further into the York and Shackleton, the latter as a long-range maritime patrol aircraft. The Shackleton would remain in service until 1991.

Opposite / A line of Gloster Meteors at RAF Manston in January 1945. The Meteor, the first Allied jet fighter, remained in service until the early 1980s.

Above / Meteors of 208 Squadron loop in formation near Abu Sueir air base in Egypt in 1952. Meteors would later take part in the abortive mission over the Suez Canal in 1956.

Canberra B.6s of Nos 109 and 139 Squadrons at Binbrook, Lincolnshire, 1957.

THE FARNBOROUGH
AIR SHOW

In the late 1940s and 50s, the old RAE Farnborough was home to the greatest show on Earth. Every September, as many as 160,000 people would gather to watch the latest aircraft emerging from British aircraft factories. Above them, jet aircraft of unbridled innovation and technological achievement were being flown by daredevil test pilots. Everything in the air was British – no other nation was allowed to display there; Britain had a world-class, world-beating air industry and Farnborough was its show-case. This was where De Havilland, Avro, Vickers, Hawker, Gloster and a host of others who provided the RAF with the latest aircraft showed off their wares to potential overseas buyers and to a wide-eyed British public. The six-day show would attract foreign officials and government ministers, heads of overseas air forces, and a host of other dignitaries. Even the Soviets would attend.

The weekend was when the show opened to the public. For many, Farnborough was intoxicating. On the ground, Britain may still have looked a little drab and war-weary, but in the skies above, these planes, with their ultra-modern, futuristic designs were breaking world speed and altitude records. Those crowds believed they were looking at the future – and more specifically, a future that promised for Britain both wealth and power.

Aircraft design had changed dramatically since the start of the Second World War: new jets were much faster, louder and looked so very different too. The test-pilots that flew these pioneering new aircraft were household names: men like John Cunningham, who had been a highly successful night-fighter during the war, or Neville Duke, a leading wartime fighter pilot, were as famous as sports stars to many. Needless to say, theirs was an incredibly

dangerous job, but all understood that throwing these new jets around the sky with barnstorming displays of speed and agility was the best way to sell them to other countries. In effect, they had become salesmen and on the back of their skill and courage there was now much at stake.

At the 1952 air show, De Havilland were showcasing their new DH110 jet, which was expected to break the sound barrier. For five days, it wowed crowds with its astonishing grace, agility and, above all, speed, for the manoeuvre that everyone wanted to see was the pilot, John Derry, sweeping down and hurtling past the flight line and, with a deafening sonic boom, breaking the magical sound barrier. Derry had been a successful wartime fighter pilot and had commanded a Typhoon squadron in north-west Europe. Softly-spoken and quiet, he was not a flamboyant type, but had a well-deserved reputation as a superb pilot and post-war had been quickly brought into De Havilland's fold. On the sixth day of the show, a fault had developed in the DH110 he had been flying so far at the show and so he had headed back to De Havilland's headquarters at Hatfield to pick up a second jet.

On cue, he dived down out of the sky, the crowd heard the triple bang of the sonic boom, Derry hurtled past

the crowd and climbed – and then the aircraft simply disintegrated. Derry and his navigator were killed and the two jet engines plummeted into the crowd, killing a further 29 people and injuring a further 80. Incredibly, just 20 minutes later, Derry's great friend Neville Duke took off in his Hawker Hunter and broke the sound barrier again. The show had to go on; there was too much at stake for any thought of cancellation. Captain Eric "Winkle" Brown had been a Fleet Air Arm pilot in the war, then a leading test pilot. "I know we worried a little bit at Farnborough," he said, "that some of the displays that were given were a bit near the knuckle. And if you'd prepared your display and you saw someone that put on a real cracker, you often thought to yourself, I've got to up my ante a bit here. And then you'd suddenly put on a bit of a show you hadn't rehearsed and that was a grave danger."

Farnborough continued to wow the crowds, however, with the chance to glimpse one of the new prototypes eagerly anticipated. In 1954, Avro test pilot Roly Falk, wearing pinstripe suit and tie, took off in the giant new Vulcan V-Bomber and promptly barrel-rolled it – a stunt in a bomber that was previously unheard of.

The war may have been over, but a new danger to the world's future security soon reared its head. Even before the end of the war, it was clear there were fundamental differences of political and ideological outlook between the Western Allies and the Soviet Union. The wartime alliance had always been one of necessity and with the war over it did not take long for co-operation in the post-war world to break down. In a speech in the US in March 1946, the former Prime Minister, Winston Churchill, talked of an "iron curtain'" of Soviet-backed communism that had descended across Eastern Europe. Certainly, Europe was now divided and became only more so as the Soviet Union backed a communist take-over of Czechoslovakia in February 1948. Berlin also lay in what was now communist East Germany, but the city itself was split into four different sectors, three controlled by Britain, France and the USA, and a fourth by the USSR. In June 1948, in a bid to gain control of the whole of Berlin, Stalin blocked all rail, road and canal access into West Berlin.

This marked the start of the Cold War. Refusing to be cowed by Soviet aggression, the Allies began flying in food, fuel and supplies to the two million beleaguered West Berliners. Three air corridors were developed and the RAF was called upon to use its air transport fleet of Dakotas and the new Hastings and Yorks, both of which could carry almost ten tons of supplies. An entirely successful operation, the Berlin Airlift continued until May 1949 when the Soviets conceded and lifted the blockade. By its end, the RAF effort had delivered an extraordinary 394,509 tons and 65,857 sorties.

During the Berlin operations, the USAF had taken over three British air bases. The initial agreement had been for 30 days, which was then extended for 60, and then for the duration of the entire Berlin Airlift. Two years on, the Americans remained in Britain and their nuclear strike force had become a permanent presence on British soil. This meant the USA was capable of launching a nuclear attack on the Soviet Union from British air bases and without any prior consultation with Britain. "We must not forget that by creating the American bases in East Anglia," Winston Churchill warned in 1951, "we made ourselves the target and perhaps the bullseye of a Soviet attack." By this time, the USSR also had a nuclear capability; much to the shock of the West, the Soviets successfully tested their first nuclear bomb in the desert of Kazakhstan on 29 August 1949. However, they still did not have the capability of delivering a nuclear warhead to the United States. They could, however, reach Britain. The worry for the British

Right / A Short Sunderland flying boat on Lake Havel in Berlin during the Berlin Airlift. Only when the lake iced up in December were the Sunderlands withdrawn from the operation.

was that the Americans might deliver a pre-emptive strike on the USSR, and that the United Kingdom would then become the target for any retaliation. A climate of fear and mutual suspicion had developed between the former wartime allies, with Britain in the middle.

Britain and the United States had never been formal allies, but rather had been wartime coalition partners. This had included much sharing of science and technology, and British scientists had played a vital role in the Manhattan Project, which developed the first atomic bomb. Post-war, however, the McMahon Act of 1946 had blocked any further co-operation with the British over nuclear research. This meant that Britain faced a stark choice: either to develop its own nuclear programme or face strategic and tactical impotence. No matter the cost, it was agreed in 1946 that Britain did need its own nuclear capability and in October 1952 the fruits of that research and development were a thermonuclear bomb successfully tested off the coast of Australia. Britain had become the third nuclear power and national security had unquestionably significantly improved as a result.

It was also accepted that the RAF would become the service responsible for delivering this nuclear capability. Contracts for a new jet strike force had been issued in 1946 and the V-Force of three nuclear bombers, the Victor, Valiant and Vulcan was unveiled in 1953 and became operational as the Bomber Command Main Force in 1955, when the Valiant became the first of the V-Bombers to enter service.

"The British Government is determined to remain in Berlin under all circumstances."

„Die britische Regierung ist entschlossen, unter allen Umständen in Berlin zu bleiben."

Mr. BEVIN, British Foreign Minister
May 20th, 1948   Scarborough

PSS (B) 7466/10m/5.48

Opposite, above / German workmen unload supplies from York aircraft that have just landed at Gatow airfield during the 1948 Berlin Airlift.

Opposite, below / An extract from a statement by British Foreign Minister Ernest Bevin expresses Britain's determination not to be pushed out of Berlin by the Soviet blockade.

Above, top / Vulcan, Victor and Valiant bombers, the trio that made up the RAF's "V-Force" fly past during 20th-anniversary commemorations of the Battle of Britain in 1960.

Above / British and US aircrews monitor a board showing the progress of the Berlin Airlift. In total, the RAF flew over 175,000 sorties during the operation.

THE
V-BOMBERS

The British Government realized that it was no use having a nuclear arsenal if the bombs could not be successfully delivered to their targets. The age of propeller-driven bombers was over; what was needed was a new high-altitude strike force that could rely on height and speed to defend itself. Tender B35/46, was issued in 1946 to meet the Air Staff's requirement for a new jet bomber capable of flying 6,100 kilometres (3,800 miles) and at least 930 km/h (580 mph). It also needed to be able to carry an atomic bomb of up to five tons and be capable of dropping it from 15,000 metres (50,000 feet), more than double the normal height from which Lancasters operated.

Six companies pitched for the tender and two were accepted. The first was from Avro, who had built the Lancaster, and had bid with a radical new design. The Avro Special Projects Department – known as the Avro Babes – borrowed a German glider design that they had seen on a post-war scouting trip of German aviation factories, and which used a delta wing. They believed a similar design would provide the required stability, lift, and strength at high altitude and so built a first scaled-down prototype. This revealed buffeting at high speed, which was swiftly resolved by adding a kink into the leading edge of the wing. The eventual Vulcan bomber

was huge. Powered initially by four Rolls-Royce Avon engines, the Vulcan used enough sheet metal to cover one-and-a-half football pitches and including three kilometres (two miles) of tubing and 22.5 kilometres (14 miles) of electrical cabling. The new bomber was unveiled on 30 August 1952 and was flown by Roly Falk, a wartime test pilot who had specialized in flying captured German aircraft. Taking off from Woodford, it brought the nearby city of Manchester to a halt as people looked to the skies to see this astonishing giant vision from the future.

Meanwhile, the second bomber was being designed and developed by Handley Page, who had built the wartime Halifax bomber. The Victor, as it would be called, was every bit as futuristic as the Vulcan, with a crescent wing and superb aerodynamic curves and ultra-modern cockpit shape. Fast, powerful and strong, it was a wonderful new-age jet bomber and the most electronically and aerodynamically advanced bomber the world had ever seen. "This was a predator of the skies," said Victor pilot, Nigel Fountain. "To this day, the Victor

is one of the most evil-looking aircraft. If you wanted a plane that looked like something out of 1950s science fiction, it was the Victor."

While the Vulcan and Victor were being developed, Vickers Armstrong entered the ring with a third design, having promised the Air Ministry they could build a bomber both on budget and on time and meet the required criteria. With mounting apprehension, that time was ticking and the Vulcan and Victor were falling behind schedule, Vickers were given the go-ahead. Their Valiant, less futuristic than the other two, first flew in 1951 and was in full production two years later. "I loved it the first time I flew it," said Peter West of the Valiant. "We went very fast and very high. That was brilliant." ➤

Opposite / Ground crews clear the area around a Vulcan bomber on 24 June 1958 as Prince Philip is about to take a flight in the aircraft after inspecting the V-Force at RAF Wyton.

Above / Handley Page Victor B.1s at RAF Cottesmore in 1958. After the delivery of the nuclear deterrent was transferred to Polaris nuclear submarines in 1959, the Victors were used as aerial refuelling tankers.

Pilots were equally effusive in their praise of the Vulcan. "What a beautiful aeroplane," said pilot Flight Lieutenant Don Exley. "I used to say it was the best fighter I ever flew. More highly manoeuvrable." Britain now had a modern jet bomber force to rival any in the world, as later proved by bombing competitions in the United States known as "Red Flags". Simulating nuclear warfare, they would fly target runs over American cities. "They were very impressed with our planes," says Exley. "Theirs took a long time to take off when fully loaded. They would take off and fly straight without turning, but we could turn almost immediately. This difference in aeroplanes was quite marked."

In the space of just one decade, air power had dramatically changed. Vast bomber fleets would never appear over the skies again. Instead, just a handful of high-altitude bombers could deliver more destruction than all the RAF's wartime Lancasters put together. "I firmly believed in the deterrent," says V-Bomber Air Electronics Officer Peter West. "None of us wanted to drop it. But we all believed this was the only way to keep peace. Our existence meant there would be no war."

Above / A Vickers Valiant at RAF Marham, Norfolk, in 1964. The Valiants were retired the following year after dangerous stress fractures were discovered.

While escalation of the Cold War remained the greatest threat to national security, Britain still had a large overseas empire and accompanying obligations. The biggest of these commitments was in Malaya. During the Japanese occupation in the war, the Allies had supplied arms, equipment and also men to help the anti-Japanese resistance led by the Malayan Communist Party (MCP). The MCP controlled an army of insurgents called the Malayan Races Liberation Army (MRLA), although some 95 per cent of the insurgents were ethnic Chinese, and by 1948 had begun turning on their former allies with a campaign of attacks on rubber plantations, tin mines and both British and Malay civilians. Although Malayan independence beckoned, Britain still supported a long 12-year campaign (the Malayan Emergency) to bring both the MCP and MRLA to heel. Operation FIREDOG was launched to oust the communists and consisted almost entirely of jungle operations. It was principally the task of British and Commonwealth ground troops to support the Malayan Police, but as in north-east India and Burma during the war, air power was critical both in providing vital reconnaissance work and also in delivering supply drops both to troops and villagers. Avro Lincolns of the RAF's Far East Air Force played a big part, but so did small Austers, used as eyes in the air and ideal for such operations because they could land and take off on a 140-metre (150-yard) strip cleared from the jungle.

Helicopters were also used and the Casualty Evacuation Flight was formed at RAF Seletar in Singapore in April 1950. Among the first pilots in the flight was Flight Lieutenant John Dowling; their first casualty was a Malay policeman who had been wounded in the leg. "Theoretically, it had no other function," said Dowling, "but of course that didn't last and soon we were doing other emergency evacuations too." It was not long before they were delivering troops into the battle zone, too. Dowling and his colleagues were involved in the first mass-landing assault in a mangrove swamp near Kuala Lumpur and later began regularly transporting the SAS who had been reformed for the Emergency. Three years later, in February 1953, 194 Squadron became the RAF's first purely rotary – helicopter – squadron and over the following years would fly more than 35,000 sorties as part of the Malayan Emergency.

Malaya gained its independence in 1957, by which time the rebels had been all but defeated. Even so, the Emergency was not officially over for a further three years. During this longest British campaign since the Napoleonic War, Britain and the RAF had played a key role in the eventual defeat of the communists.

Above / A Westland Whirlwind helicopter lands men of the RAF regiment in Selangor State in April 1957, during the Malayan Emergency.

During that same period there were other crises in which British forces played a part. In June 1950, North Korean forces invaded South Korea and the United Nations, led by the United States, entered the war on behalf of the South. Britain's involvement was limited because of the ongoing Emergency in Malaya, but ground troops, naval forces and the RAF were all sent and played an important part in the war. In the air, both jets and piston-powered aircraft tussled over the skies and US F-84 Thunderjets and F-86 Sabres found themselves confronting MiG-15 Soviet-built jets of the Chinese Air Force. That these jets were flying at all was largely down to the woeful decision by the British government at the end of the war to sell the Soviet Union an engine on the understanding it would be used for commercial purposes only. The Soviets reverse-engineered the Nene far more quickly and efficiently than the British had imagined and promptly put them in their new MiG fighter-jet airframes.

A handful of RAF pilots were, however, seconded to the USAF, including Flight Lieutenant John Nicholls. On 8 December 1952, Nicholls was flying his 99th operational sortie in an F-86 Sabre when he finally managed to get on the tail of a MiG-15 he was chasing and saw it taking hits. The MiG pilot tried to shake him off by diving in a spiral. "I went down after him," recalled Nicholls, "as fast as I could, and that was really very fast." At such transonic speeds close to Mach 1, keeping control of the jet was not easy and he had some difficulty getting his guns back on target. "I did it fleetingly," he added, and each time I did I pulled the trigger." He could see he was hitting his opponent, but then Nicholls' gunsight stopped working. Lower and lower they flew, Nicholls determined to get his man. Eventually, flames appeared from the MiG, and just as Nicholls turned for home, he saw it flick and go straight into the ground. The next day, Nicholls flew his 100th and final mission over Korea.

The RAF contribution in Korea was limited because, although by the mid-fifties one in six British scientists was working in the aviation industry and some quarter of a million people were working in British aircraft factories, defence was becoming ever-more expensive and the creation of nuclear warheads and the V-Force was absorbing much of the Air Ministry's budget. Britain simply could no longer afford a vast empire and all the defence commitments that this incurred. For four years from 1952, Britain faced a further emergency in Kenya, in what became known as the Mau Mau Rebellion. Reconnaissance and supply drops were also the RAF's prime role there, although ground-attack De Havilland Vampire jets from Aden were also sent to support efforts to end the rebellion on the ground.

Above / An Avro Lincoln of 61 Squadron flies past Mount Kenya in April 1954 during counter-terrorist operations to put down the Mau Mau Uprising.

Opposite / De Havilland Vampires on a training mission in Malta in 1954. The aircraft were used in the Malayan Emergency and Mau Mau Uprising before being superseded by the Venom in the mid-1950s

Then, in 1956, came the Suez Crisis. Two years earlier, the Egyptian Government ended the treaty that allowed Britain to operate bases in the Suez Canal Zone. Soon after the withdrawal had been completed, Egypt announced plans to nationalize the Suez Canal. War between Israel and Egypt looked increasingly likely, while Egypt was receiving increasing arms and support from the Soviet bloc. France and Britain, who both wanted to maintain freedom of movement through the canal, began concentrating forces in the Mediterranean on Malta and Cyprus while diplomatic efforts were made to resolve the crisis. When this failed, Operation MUSKETEER dropped large Anglo-French airborne forces into Egypt to occupy the Canal Zone. Meanwhile, the RAF's main task was to neutralize the Egyptian Air Force and their Soviet-built MiG-15s and Il-28 bombers. Canberra jet bombers and the new Valiants carried out a number of high-level precision raids, while RAF Venoms and French fighter-bombers launched a series of low-level ground attacks. RAF operations were successful and achieved all their aims, but the brief war ended just seven days later with a UN-backed diplomatic solution which saw the United Nations take over control of the Canal Zone. It was a diplomatic humiliation for both Britain and France, which saw the resignation of the Prime Minister, Anthony Eden, and marked the end of Britain's independent imperial dominance. The end of the British empire was nigh, the world was changing dramatically fast, and the two biggest superpowers in the world were now the USSR and the USA; Britain as a great power was in decline. In this new era, the RAF would need to evolve too.

Above / Ground crew load bombs into a Canberra of 101 Squadron in October 1956 during operations against Egyptian air bases as part of the Suez Crisis.

Opposite / A Canberra is armed at RAF Hal Far, Malta, in 1956 prior to its mission along the Suez Canal. The RAF's attacks on a dozen airfields decimated the Egyptian air force.

Opposite / One of the most beautiful jet aircraft ever designed: the superb but ill-fated Fairey Delta 2.

9

THE COLD WAR –
THE
AGE OF
DÉTENTE

In April 1949, the North Atlantic Treaty Organization (NATO), had been formed, binding the western powers, including the USA and Britain, in alliance against the mounting threat posed by the Soviet Union. NATO, claimed its first Secretary-General, Sir Hastings Ismay, was designed "to keep the Russians out, the Americans in, and the Germans down." It was a line that until 1989 remained a pithy shorthand to describe the alliance's aims. However, if NATO was a grand alliance of the West, then the Warsaw Pact (established in 1955) was the same in the East, and bound the USSR and its satellite Eastern European states together behind the Iron Curtain.

Britain was utterly committed to playing a major role in NATO and by 1952, with its nuclear programme in place and the V-Force of new jet bombers in development, defence spending reached a colossal 11.8 per cent of GDP. This was not a sustainable commitment, however, and maintaining the UK's very high military budget required a delicate financial juggling act. If NATO and the Cold War were to be Britain's defence priority, that meant, in the uncertain post-war economy, that there would have to be a gradual reduction of overseas commitments, equipment and manpower.

Perhaps one of the most significant turning points for the RAF was the 1957 White Paper on Defence drawn up by Duncan Sandys, the Minister of Defence and coincidentally Winston Churchill's

son-in-law. Sandys had been a gunner in the war, had commanded a missile testing battery in Wales and also been in London during the V-2 rocket raids. These experiences left a lasting impression on Sandys and in his vision of future war, he was convinced that guided missiles, particularly surface-to-air missiles, threatened all aircraft, and that as a result the RAF would eventually become redundant. Defence spending, he argued, and particularly the RAF's allocation, should reflect this changing technology.

The main aim of the White Paper, however, was to stem the colossal over-spending on defence, and so in the short-term he suggested a number of measures to cut costs. Among these was the amalgamation of many of Britain's aircraft manufacturers. English Electric, Vickers-Armstrong, Bristol and Hunting were all merged to form the British Aircraft Corporation, while de Havilland, Folland and Blackburn were to merge with Hawker Siddeley, which would retain its name. In return, contracts for new aircraft and equipment were offered as compensation. These, however, were to be fewer in number and of fewer varieties.

Cutting the number of aircraft was a main feature of Sandys' White Paper. The cull was to be brutal. The beautiful Fairey Delta 2, currently holding the world speed record, was scrapped and Fairey forced to merge with Westland whose focus now was on helicopter production. Hawker Siddeley were not immune either. They had already spent £1 million – then a huge sum – on their supersonic interceptor, but that was another project cancelled by the Government. Saunders-Roe's dazzling new jet rocket fighter was also chopped, even though both the RAF and Admiralty were keen to bring it into service, while Avro also had two future jet fighters shelved. All these projects had taken years of immense dedication and skill and the result was a number of aircraft of

Left / Duncan Sandys, the British Defence Minister and author of the 1957 White Paper that led to draconian cuts to the RAF, pictured at the Houses of Parliament.

Opposite / The 1957 government White Paper on Defence; its assumption that missiles would supersede aircraft led to rapid consolidation in the British military aviation industry.

DEFENCE

Outline of Future Policy

*Presented by the Minister of Defence to Parliament
by Command of Her Majesty
April 1957*

LONDON
HER MAJESTY'S STATIONERY OFFICE
NINEPENCE NET

Cmnd. 124

---

2

7. Over the last five years, defence has on an average absorbed 10 per cent. of Britain's gross national product. Some 7 per cent. of the working population are either in the Services or supporting them. One-eighth of the output of the metal-using industries, upon which the export trade so largely depends, is devoted to defence. An undue proportion of qualified scientists and engineers are engaged on military work. In addition, the retention of such large forces abroad gives rise to heavy charges which place a severe strain upon the balance of payments.

Britain's Responsibilities

8. A defence plan, if it is to be effective and economical, must be based on a clear understanding of the military responsibilities to be discharged. Britain's armed forces must be capable of performing two main tasks:—

(i) to play their part with the forces of Allied countries in deterring and resisting aggression;
(ii) to defend British colonies and protected territories against local attack, and undertake limited operations in overseas emergencies.

The aim must be to provide well-equipped forces sufficient to carry out these duties, while making no greater demands than are absolutely necessary upon manpower, money and other national resources.

9. Frequent changes in defence policy are wasteful and disturbing. Experience has shown that the rapid progress of scientific development and fluctuations in the international situation make it difficult to foresee future military requirements with any certainty, and that consequently a good deal of flexibility must be maintained. Nevertheless, an attempt must be made to establish a broad framework within which long-term planning can proceed.

Collective Defence

10. The growth in the power of weapons of mass destruction has emphasised the fact that no country can any longer protect itself in isolation. The defence of Britain is possible only as part of the collective defence of the free world. This conception of collective defence is the basis of the North Atlantic, South-East Asia and Baghdad alliances.

11. The trend is towards the creation of integrated allied forces. Therefore, provided each member nation plays its fair part in the joint effort, it is not necessarily desirable that each should seek to contribute national forces which are by themselves self-sufficient and balanced in all respects. But whatever yardstick is taken, it is impossible to escape the conclusion that Britain has been bearing a disproportionately large share of the total burden of Western defence. Moreover, in assessing the value of her military effort, it must be remembered that, apart from the United States, Britain alone makes a contribution to the nuclear deterrent power upon which the peace of the world so largely rests.

Nuclear Deterrent

12. It must be frankly recognised that there is at present no means of providing adequate protection for the people of this country against the consequences of an attack with nuclear weapons. Though, in the event of war, the fighter aircraft of the Royal Air Force would unquestionably

---

be able to take a heavy toll of enemy bombers, a proportion would inevitably get through. Even if it were only a dozen, they could with megaton bombs inflict widespread devastation.

13. This makes it more than ever clear that the overriding consideration in all military planning must be to prevent war rather than to prepare for it.

14. While comprehensive disarmament remains among the foremost objectives of British foreign policy, it is unhappily true that, pending international agreement, the only existing safeguard against major aggression is the power to threaten retaliation with nuclear weapons.

15. The free world is to-day mainly dependent for its protection upon the nuclear capacity of the United States. While Britain cannot by comparison make more than a modest contribution, there is a wide measure of agreement that she must possess an appreciable element of nuclear deterrent power of her own. British atomic bombs are already in steady production and the Royal Air Force holds a substantial number of them. A British megaton weapon has now been developed. This will shortly be tested and thereafter a stock will be manufactured.

16. The means of delivering these weapons is provided at present by medium bombers of the V-class, whose performance in speed and altitude is comparable to that of any bomber aircraft now in service in any other country. It is the intention that these should be supplemented by ballistic rockets. Agreement in principle has recently been reached with the United States Government for the supply of some medium-range missiles of this type.

Defence of the Deterrent

17. Since peace so largely depends upon the deterrent fear of nuclear retaliation, it is essential that a would-be aggressor should not be allowed to think he could readily knock out the bomber bases in Britain before their aircraft could take off from them. The defence of the bomber airfields is therefore an essential part of the deterrent and is a feasible task. A manned fighter force, smaller than at present but adequate for this limited purpose, will be maintained and will progressively be equipped with air-to-air guided missiles. Fighter aircraft will in due course be replaced by a ground-to-air guided missile system.

Civil Defence

18. While available resources should as far as possible be concentrated on building up an active deterrent power, it would be wrong not to take some precautions to minimise the effects of nuclear attack, should the deterrent fail to prevent war. Civil Defence must accordingly play an essential part in the defence plan.

19. As in other fields, the country's economic capacity limits the effort which can be devoted to this purpose. In 1957/58 the main task will be to keep the existing local organisation in being, so as to provide a basis on which realistic planning can continue. The necessary training equipment will be provided. Essential research will proceed; and work on emergency communications and on setting up the fall-out warning and monitoring system will go on. These preparations will provide a framework for expansion, should that later be necessary.

Europe and Atlantic

20. The possession of nuclear air power is not by itself a complete deterrent. The frontiers of the free world, particularly in Europe, must be firmly defended on the ground. For only in this way can it be made clear that aggression will be resisted.

21. Britain must provide her fair share of the armed forces needed for this purpose. However, she cannot any longer continue to make a disproportionately large contribution.

22. Accordingly, Her Majesty's Government, after consultation with the Allied Governments in the North Atlantic Council and in the Council of the Western European Union, have felt it necessary to make reductions in the British land and air forces on the Continent. The strength of the British Army of the Rhine will be reduced from about 77,000 to about 64,000 during the next twelve months; and, subject to consultation with the Allied Governments in the autumn, further reductions will be made thereafter. The force will be reorganised in such a way as to increase the proportion of fighting units; and atomic rocket artillery will be introduced which will greatly augment their fire-power.

23. The aircraft of the Second Tactical Air Force in Germany will be reduced to about half their present number by the end of March, 1958. This reduction will be offset by the fact that some of the squadrons will be provided with atomic bombs. A similar reduction will be made in the light bomber force in England, which is assigned to NATO.

24. The rôle of naval forces in total war is somewhat uncertain. It may well be that the initial nuclear bombardment and counter-bombardment by aircraft or rockets would be so crippling as to bring the war to an end within a few weeks or even days, in which case naval operations would not play any significant part. On the other hand, there is the possibility that the nuclear battle might not prove immediately decisive; and in that event it would be of great importance to defend Atlantic communications against submarine attack. It is therefore necessary for NATO to maintain substantial naval forces and maritime air units. Britain must make her contribution, though, for the reasons explained earlier, it will have to be on a somewhat reduced scale.

Middle East

25. Outside the area covered by the North Atlantic Alliance, Britain has military responsibilities in other parts of the world, in particular in the Middle East and South-East Asia.

26. Apart from its own importance, the Middle East guards the right flank of NATO and is the gateway to the African continent. In the Arabian Peninsula, Britain must at all times be ready to defend Aden Colony and Protectorates and the territories on the Persian Gulf for whose defence she is responsible. For this task, land, air and sea forces have to be maintained in that area and in East Africa.

27. In addition, Britain has undertaken in the Baghdad Pact to co-operate with the other signatory States for security and defence, and for the prevention of Communist encroachment and infiltration. In the event of emergency, British forces in the Middle East area would be made available to support the Alliance. These would include bomber squadrons based in Cyprus capable of delivering nuclear weapons.

THE FAIREY DELTA 2

Left / A Fairey Delta pictured at the Farnborough Air Show in 1958. The Delta's high-profile aviation record-breaking had failed to save it from Duncan Sandys's axe the year before.

There is a rule of thumb that claims if an aircraft looks right, then usually it will fly well too, and it is amazing how often that holds true. The Fairey Delta 2 has to be one of the most beautiful aircraft ever built: sleek, elegant, with a long feline nose that could be hydraulically hinged down to give the pilot a better view during taxiing and take-off, and neat delta wings. Incredibly, the Delta 2 had originally been designed in the late 1940s but it still looks modern even to this day. Fairey were a small company and not awarded a contract until 1950, while construction only began two years later. The test pilot was Peter Twiss, a former Fleet Air Arm pilot, and he was soon reporting that it flew every bit as well as its stunning design promised. On its fourteenth test flight in November 1954, however, Twiss had only narrowly avoided being killed when a faulty gasket had deprived the engine of fuel. The prototype was rectified and Twiss later returned to the controls. "If ever the Delta had had to justify herself to me after her crash," noted Twiss, "this wonderful little aircraft did so that morning when she flew as gently as a bird into the hard supersonic October sky. From that moment, I knew we had a world-beater." Sure enough, the Delta 2 effortlessly smashed the world speed record of 1,323.30 km/h (822.26 m.p.h.) held by America's Colonel "Dude" Hanes in a Super Sabre: in March 1956, Twiss recorded a speed of 1,822 km/h (1,132 m.p.h.).

With this impressive record under their belts, both Fairey and Twiss were confident the Delta 2 could become a front-line supersonic jet fighter for years to come. Certainly, senior officers with the RAF were giving them as much encouragement as they could. Sadly, supersonic flights under 9,100 metres (30,000 feet) over populated areas had recently been banned in Britain, and nor could they find anyone to insure them against damage claims caused by the Delta 2's sonic booms. As a result, Fairey and Twiss were forced to continue testing in France instead, where the Dassault company agreed to let them base themselves at their headquarters at Cazaux, south of Bordeaux.

Despite its brilliance, the Delta 2 was summarily axed by Duncan Sandys in 1957, an act of aviation vandalism that still causes pain to enthusiasts even to this day. One survives and is on display at the RAF Museum in Cosford. Walking around it inspires mixed thoughts of wonder, awe, and sadness. It is an aircraft that should be hurtling across the sky, creating sonic booms, not standing stationary in a museum hangar. Ironically, some of the Delta 2's features were later absorbed by Marcel Dassault into his superb and long-serving Mirage jet fighter; certainly the early Mirage looked very similar to the Delta 2. "If it were not for the clumsy way in which you tackle things in Britain," Dassault once said, "you could have made the Mirage yourselves."

awesome power and brilliance, most of which were chopped up and destroyed before the very eyes of those who had worked so hard to create them.

Not surprisingly, the Sandys White Paper was not well received by the RAF, whom it threatened the most, nor by Britain's aviation industry as a whole. British air power had been in the ascendancy for more than 20 years and had played a significant part in the Allied victory in the Second World War. Since the war's end, British jet aircraft had been world leaders, and not just the military aircraft but commercial, too. The de Havilland Comet had been the much-vaunted first commercial jet airliner (and perhaps the most beautiful ever built). It was an aircraft that had promised to dominate the world, but then disaster had struck. First one, then a second and then a third crashed. The fault, it seemed, was due to stress fractures developing in one of the cockpit windows. Large orders were cancelled and by the time the problem had been resolved, de Havilland had lost out to a new generation of Boeing jet airliners and would never regain its primacy.

Britain's empire of the skies had briefly dazzled, but by 1957 was, like Britain as a whole, no longer a pre-eminent global power. The Sandys White Paper ripped out the soul of Britain's aviation industry and it would never recover. At the time, many recognized the report as strategic hubris, and so it was to prove. Surface-to-air missiles would certainly play a greater role in the future, but there was still an important function for jet aircraft as the next few decades would demonstrate. The RAF's prime function in the Cold War was to provide a deterrent and to prevent any escalation into all-out war between East and West. Fast jets provided muscle: to intercept Russian incursions into air space, and to demonstrate that British air power was technologically the equal, if not superior to that of the Soviet Union. Supersonic jets showed the world that the RAF was still a major force with which to be reckoned. The White Paper cull was, and remains, both ill-judged and a gross travesty.

Perhaps more justifiable was the reduction of the RAF's overseas presence around the world. National Service was also scrapped in 1960, and by 1964 the RAF had once again returned to being an all-volunteer force. The Air Ministry and Admiralty were also merged with the War Office to become the single, tri-service Ministry of Defence, but while there is no doubt the Sandys report led to a more streamlined RAF, this was achieved by surrendering independence and capability.

Above / A de Havilland Comet C.4 of 216 Squadron in November 1971. Although a series of high-profile crashes doomed the Comet as a civilian airliner, it remained in service as a troop carrier.

Nonetheless, certain aircraft did survive, including the English Electric Lightning, which, as it happened, had been test flown by former fighter pilot Roland "Bee" Beamont the very day Sandys published his White Paper. Even so, Sandys spared it only begrudgingly. A beast of an aircraft, it was about as thirsty for fuel as a jet aircraft could be, but was also immensely powerful and manoeuvrable and could reach supersonic speeds vertically. Its rate of climb was phenomenal: a Lightning could soar to 11,000 metres (36,000 feet) in under three minutes. Its shortcoming was its lack of range, although its primary role was to protect V-Force bomber airfields, for which it was very well suited. The Lighting entered service in 1960.

English Electric also had the Canberra in service, again test-flown by Bee Beamont. Developed to replace the Mosquito, it emerged as an outstanding high altitude bomber, in addition to its capability of flying at high speeds. In February 1951, one set a world record as the first jet aircraft to make a non-stop transatlantic flight. In its day, the Canberra could fly at a higher altitude than any competitor aircraft in the world. In 1957, this versatile jet established a world altitude record of 21,430 metres (70,310 feet). Both factors led to highly successful export sales. In addition to serving as a tactical nuclear strike aircraft, the type also operated as a conventional bomber and photographic and electronic reconnaissance platform. Its adaptability was reflected in its long service life: incredibly, the last Canberra was not retired until July 2006.

Above, top / English Electric Canberras fly over Aden in early 1955. The adaptable Canberras formed the backbone of the RAF's bomber and reconnaissance squadrons in the 1950s and 1960s.

Above / A Hurricane pilot in both the Battles of France and Britain, Roland "Bee" Beamont became one of Britain's top test pilots, carrying out the maiden flights of the Canberra, Lightning and TSR-2.

Opposite / A pair of English Electric Lightnings in flight. Their high performance was never matched by support from the government, which nearly scrapped them in the 1957 Defence review.

Despite Sandys' prediction about the increasing redundancy of jets, the V-Force would remain in service – for the time being at any rate. The nuclear strike force squadrons existed on round-the-clock readiness called QRA – Quick Reaction Alert. By the early 1960s, tensions between East and West were at breaking point and over the weekend of 27 and 28 October 1962, when Soviet missile sites were discovered on Cuba. Britain – and the world – was on the edge of Armageddon as President John F. Kennedy and Soviet leader Nikita Khrushchev squared up to each other in the ultimate game of brinkmanship. Although war was averted, the QRA squadrons were ready to be scrambled and drop nuclear bombs over the Soviet Union. Each V-Bomber was equipped with the Mass Start-Up Facility, which enabled all four engines in the Vulcan, for example, to be fired up simultaneously. Readiness for the crew was 15 minutes: the time it would take for a Soviet nuclear missile to reach Britain. "The whole of the V-Force were there for a one-shot mission," said Rip Kirkby, a former V-Force pilot. "The one mission you were going to fly for real was the one you wouldn't have come back from."

Already, though, the V-Force's days were numbered. In 1960, US pilot Gary Powers was flying a U-2 spy plane over the Soviet Union at 18,000 metres (60,000 feet) when he was shot down by a Soviet surface-to-air missile. It was evidence that the USSR had significantly improved their arsenal, but it also meant high-altitude V-bombers were no longer impregnable. The immediate alternative was to fly low instead. "So now we were under the Russian radar," recalled Peter West, "but we used a lot more fuel at low-level and to drop a nuclear bomb at that height is pretty much a kamikaze operation." The flash white colour scheme of the V-Force was abandoned in favour of traditional grey and green camouflage, but it soon became clear that neither the Victor nor the Valiant were effective at low-level. That left the Vulcan as the primary deliverer of the nuclear deterrent.

Nonetheless, the writing was now on the wall for the V-Force, and through the sixties, its numbers were reduced, while efforts were increasingly made to fulfill Sandys' vision of a defence system dominated by surface-to-air missiles. The first such missile system to be developed was Avro's Blue Steel, which was to be an air-

launched, nuclear-tipped, medium-range ballistic missile, and which was expected to be ready by 1957. De Havilland had also developed the medium-range Blue Streak missile – a static, ground-based weapon. Both projects suffered from rising costs and in 1957 the British Prime Minister, Harold Macmillan, agreed to station 60 American Thor missiles on British soil. The RAF Thor force of 20 squadrons was manned by Bomber Command crews trained in the USA. However, Thor was far from an ideal solution: the missiles remained under American control, were static, above ground, and lacked the mobility the V-Force had from their dispersed bases.

Blue Streak was cancelled in 1960, though Blue Steel continued under development. In 1960, Macmillan then decided to purchase the cheaper American-made GAM-87 Skybolt. This air-launched ballistic missile had an impressive 1,600-kilometre (1,000-mile) range that would allow the V-Force to stand-off from Soviet air defences and launch attacks that would be difficult to block. For largely technical reasons, the Americans suddenly cancelled the Skybolt project in 1962, gravely threatening the UK's independent nuclear deterrent, which in turn further challenged Anglo-American relations. The entire missile programme was in turmoil and becoming as expensive as the jet aircraft Sandys had scrapped. The solution was to redevelop Blue Steel, which finally entered service in February 1963 and which was carried by the Vulcans and Victors of V-Force.

Meanwhile, in the aftermath of the Cuban Missile Crisis, President John F. Kennedy and British Prime Minister Harold Macmillan concluded the Nassau Agreement in December 1962, which was later formalized as the Polaris Sales Agreement in 1963. This remains in force today. The agreement enshrined the purchase of US missiles and fire-control systems to serve aboard UK-built submarines, and was later extended to cover the current Trident programme. With the withdrawal of the Thor force in 1963, Blue Steel missiles and the WE177, the RAF's last nuclear bomb, provided the deterrent; in fact, the WE177 remained in Britain's arsenal until 1998.

Despite this, the main responsibility for delivering a nuclear strike was taken from the RAF in 1969 and handed to the Royal Navy, whose submarines were by then equipped with the Polaris missile system. By this time, Valiants had already been converted to become fuel tankers after air-to-air refuelling systems had been pioneered by the RAF – in July 1959, Wing Commander Michael Beetham, a former wartime bomber pilot and later Chief of the Air Staff, became the first man to pilot an aircraft non-stop from Britain to Cape Town in South Africa. Now, the Valiants were withdrawn and the Victors took over the refuelling task, while the Vulcan remained as a principal bomber.

Opposite / A Victor bomber is loaded with a Blue Steel nuclear-armed missile. Together with the Blue Streak, it was the last generation of British-manufactured missiles to deliver the nuclear deterrent.

Above / A Victor bomber is refuelled in mid-air by a Valiant tanker aircraft. After the mid-1960s the V-Bombers lost their raison d'être as a nuclear platform and were adapted for other roles.

The shadow of the Sandys White Paper continued to haunt future aircraft development, and particularly a new multi-purpose high-speed jet, the TSR-2 (Tactical-Strike-Reconnaissance). This had been developed from a specific requirement long before the Air Ministry ceased to exist and was designed to penetrate the Soviet forward battle area at low altitude (60 metres/200 feet) and high speeds of Mach 2, carrying out attacks with nuclear or conventional weapons. Conversely, it was also intended to provide very high-altitude, high-speed radar and photographic imagery, signals intelligence and reconnaissance. In September 1964, the first TSR-2 was rolled out at Boscombe Down to begin air testing; it was the most sophisticated weapon system the RAF had ever independently contemplated, incorporating new airframe, engine and avionics technologies.

On 6 April 1965, the team at Boscombe Down were preparing to fly the second prototype, with Bee Beamont once more the test pilot. Beamont was lunching at a nearby pub when reports of the morning's budget speech emerged and revealed that the TSR-2 project had been cancelled. Hurrying to the airfield in a last ditch attempt to get the jet airborne, they were refused permission to take off. The TSR-2 had been grounded for good. Bee Beamont was incensed. "The cancellation was a monstrous manoeuvre for political ends," he said later, "from which the striking power of the Royal Air Force and the world-leader potential of British military aviation technology never fully recovered."

A replacement was sought, and that was the Blackburn Buccaneer, originally designed for Royal Navy carriers. With their all-weather capability, Buccaneers were operated by the RAF instead, whilst British, German and Italian manufacturers perfected the swing-wing Multi-Role Combat Aircraft (MRCA), which came into service as the Tornado. First flown in 1974, it did not enter RAF service until 1979; however, the fact that the Tornado would possess broadly similar capabilities to the TSR-2 indicated just what an incredible piece of aeronautical engineering the doomed jet had been.

Even so, the Buccaneers proved fine aircraft and its unrivalled low-level stability meant that they consistently outperformed other

Below / The very first RAF Tornado GR1 just after its delivery to RAF Cottesmore on 30 June 1980.

strike aircraft, and they regularly did well in the annual Anglo-US "Red Flag" exercises in Nevada. The first RAF unit to receive them was 12 Squadron at Honington in October 1969, and they would remain in RAF service for an impressive 25 years, before withdrawal in 1994.

One project that did survive was Hawker Siddeley's Vertical/Short Take Off and Landing (V/STOL) technology that eventually became the Hawker Harrier. This superb aircraft had been lucky to survive, as two other V/STOL projects, the P1154 supersonic fighter and the HS681 V/STOL transport, had also been cancelled during the 1965 Defence Review. Four years on, 1969 was to prove a busy year for the RAF and its Operational Conversion Units because the arrival of the Harrier into squadron service also coincided with

the activation of the first Buccaneer squadrons and also those of the new McDonnell-Douglas Phantoms, which had been bought from the USA in another sign of the changing times. At the time, the RAF's primary interceptor remained the Lightning, which despite surviving the Sandys cull, had low range, loiter time and was under-armed. This had not been an issue while its prime role had been protecting home airfields, but now the V-Force was being phased out, its lack of range was a major hindrance in the longer operations required against Soviet reconnaissance aircraft over the North Sea and North Atlantic. Consequently, 43 Squadron based at Leuchars in Scotland was re-equipped with F-4 Phantoms and began operating alongside the Lightnings of 11 Squadron in the QRA role.

⊙

Reflecting the changing times, it was accepted that the old system of RAF commands needed reshaping. At the end of the Second World War, the RAF had comprised ten commands – Bomber, Fighter, Coastal, Transport, Balloon, Flying Training, Technical Training, Signals, Maintenance and Reserve Commands. Some of these had merged already, but in April 1968, Bomber and Fighter Commands combined to form Strike Command, with offensive and defensive roles concentrated into No 1 and 11 Groups respectively. Strike then absorbed Signals and Coastal Commands in 1969, and Air Support (formerly Transport) Command in 1972; other mergers

have followed as the RAF's streamlining has continued, reflecting fewer defence commitments and a strict adherence to the defence budget. This restructuring was also indicative of the shrinking size of the entire RAF. From 270,000 personnel in 1952, a decade later it numbered 148,000. By 1968 this had reduced further to 120,000 and by 1976, a mere 90,000 – exactly a third of the 1952 strength.

Above / A Blackburn Buccaneer of 16 Squadron flies over West Germany in the 1970s. Initially designed for the Royal Navy, the RAF adopted it after the cancellation of the TSR-2.

THE
HARRIER

The iconic Harrier jump jet began life as the Hawker P1127. Among its team of designers was the remarkable Sir Sydney Camm, responsible for many of Hawker's pre-war biplanes, as well as the Hurricane and later Typhoon, Tempest and the Hunter jet fighter. Originally intended to be a replacement for the Hunter, the P1127 was another victim of the Sandys White Paper, but out of the ashes of this abandoned project emerged the Kestrel. In 1961, Britain, the USA and West Germany all agreed to jointly purchase nine Hawker Siddeley aircraft developed from the P1127 with the aim of developing and evaluating new Vertical/Short Take-Off and Landing (VSTOL). The idea of creating enough downward thrust so that an

aircraft could rise vertically into the air had been talked about for some years, but it was Stanley Hooker and his team at Bristol Aero Engines who came up with an effective vectored thrust system in their Pegasus engine. This allowed thrust from the turbojet to be manipulated in different directions. The Kestrel was the development aircraft, but with its successful trialing, a new operational aircraft with an improved, more powerful Pegasus engine was ordered. This became the Harrier GR.1. Although subsonic, it was nimble, still fast enough, and was the first operational aircraft ever with VSTOL capabilities. The grand old man of Hawker Siddeley, Thomas Sopwith, whose Pups and Camels had tussled in the skies over the

Western Front in the First World War, was amazed by the Harrier, as were many others. "I still don't believe the Harrier," he said. "Think of the millions that have been spent on VTO in America and Russia, and quite a bit in Europe, and yet the only vertical take-off aircraft which you can call a success is the Harrier. When I saw the Harrier hovering and flying backwards under control, I reckoned I'd seen everything."

Along with the Canberra, the Harrier was only one of two post-war British aircraft to be sold to the United States, who modified it for use with the US Marine Corps. It was also adapted into the highly effective Sea Harrier, and performed brilliantly during the Falklands War, where between them, RAF and Royal Navy Harrier pilots shot down 22 confirmed enemy aircraft. It continued to see action in Iraq and Afghanistan and Harriers were much beloved by all those who worked on them and flew them. Both the Royal Navy and RAF retired their Harriers in 2011, and with their withdrawal from service so ended the service of the last entirely British-built aircraft flown by the RAF.

Opposite / A pilot carries out checks on his Harrier GR3 at Ascension Island in May 1982 on his way to join operations against Argentine forces occupying the Falkland Islands.

Below / A Harrier GR1 of 20 Squadron demonstrates the aircraft's powerful hover facility during an exercise in Germany in summer 1973.

Meanwhile, as well as its NATO commitments, the RAF was still playing a role around the world, even though the British Empire was rapidly drawing to a close and one after another former colonies flying the Union flag became independent. In 1963, the RAF had helped defend the newly created Malaysia from a sabre-rattling Indonesia. The confrontation had died down by 1966, but V-Force bombers as well as fighter jets, Canberras and helicopters had all deployed in the Malay peninsula and in North Borneo. In 1964, strike jets and Transport Command had helped quell a rebellion in the Radfan area of Aden. Although this revolt had been swiftly halted, trouble brewed up once more and three years later, the British base in Aden was evacuated. The evacuation, from May to November 1967, was the largest air lift since that into Berlin in 1948–49. There were also vital operations during the Brunei revolt in 1962–63, as well as in Borneo, along with various humanitarian operations, from Search and Rescue missions around the United Kingdom to offering relief and supplies around the world in the aftermath of natural disasters, from cyclones in Pakistan in 1970 to earthquakes in Morocco in 1960 and famines in Africa.

Above, top / A Hawker Hunter attacks Radfan rebels in Aden in 1964. The Hunter's high-explosive rockets proved potent in strikes against rebel positions.

Above/ A pilot climbs into the cockpit of his Hawker Hunter FR10 during the 1964 operation against nationalist rebels in Aden.

Opposite / A Hercules stands ready to deliver supplies during Operation KHANA CASCADE in 1973, when 2,000 tons of grain were dropped to villages in Nepal following a famine there.

Britain and the RAF went to war again in 1982 to safeguard one of the most far-flung reaches of its former empire. The Falkland Islands in the South Atlantic had remained British, although this status was disputed by Argentina. When the Argentinians invaded on 2 April that year, a task force was quickly assembled in the United Kingdom and sent to reclaim this far-off possession. Both RAF Harriers of 1 Squadron as well as naval Sea Harriers were aboard HMS *Hermes,* and both navy and RAF pilots would fight alongside one another in the ensuing ten-week battle. The RAF's Harriers focused on ground-attack missions in support of the land force, whilst the Sea Harriers provided fixed-wing air defence for the fleet; the RAF lost four and the Royal Navy six during the campaign.

The Harrier's proven flexibility in the Falklands was just one of several implications the 1982 conflict had for the RAF. The war was highly complex, being a joint air, land and sea operation, fought at a distance of 13,000 kilometres (8,000 miles) from the United Kingdom. The hub of RAF activity was Wideawake airfield on Ascension Island, roughly halfway between the UK and the war zone, and the long

Opposite / RAF GR.3s are parked alongside Royal Navy Sea Harriers on the deck of HMS *Hermes* in early May 1982, as they join operations during the Falklands War.

Above, top / Refuelling by Victor tanker aircraft based at Wideawake airfield on Ascension Island enabled Vulcan heavy bombers to operate against Argentine targets during the 1982 Falklands War.

Above / An Avro Vulcan lands on Ascension Island in May 1982. The five attacks carried out on the Falkland Islands by Vulcans were the longest-range air raids in history at that time.

distances stretched RAF air-to-air refuelling capabilities to the limit. Victor tankers bore the brunt of the work, flying over 500 sorties and enabling Nimrods to conduct maritime reconnaissance and C-130 Hercules to drop supplies to the ships of the Task Force.

Most spectacular were the Black Buck operations mounted by Vulcans to attack Port Stanley airfield and Argentinian radars on the Falklands. The Vulcan fleet was just months away from disbandment, and some aircrew had already begun conversion courses to other types. Nevertheless, enough Vulcans and aircrew were scraped together to enable five night-bombing missions to be launched. Without the support of all 13 tankers on Ascension Island, the raids would have been impossible. The first completed saw Vulcan 607, piloted by Flight Lieutenant Martin Withers, successfully bomb the airfield at Port Stanley, a trip of some 13,000 kilometres (8,000 miles). On their safe return to Ascension, Martin Withers bought his crew a round of drinks and calmly said, "Well done, guys, and thank you." It was the kind of understatement that had always marked the culture of the RAF, but there was no doubting it was a truly

extraordinary feat and immediately put the crew into the *Guinness Book of World Records* for "the longest-range attack in air history". This first and the subsequent raids that followed each took some 16 hours, and were a timely reminder that the future vision of nuclear weapons and missiles replacing aircraft, as predicted by Duncan Sandys in 1957, was wide of the mark. Wars were still to be won by the robust design of aircraft and professionalism of the crews flying them, and that continued to be the case. Chinook Bravo November, for example, has flown for more forty years, seeing service in the Falklands, Iraq and Afghanistan and is still serving in 2020.

Opposite, above / Craters left by the 21 1,000lb bombs dropped by a Vulcan bomber on Port Stanley airfield on 30 April–1 May 1982 are visible across the centre of the runway.

Opposite, below / A Vulcan B.2 bomber taxis to a halt on Ascension Island after carrying out the first of the Black Buck raids on the Falklands on 30 April–1 May 1982.

Below / British Prime Minter Margaret Thatcher speaks with the CO of the Harrier detachment at Stanley Airfield during a visit to the Falklands in early 1983.

Opposite / A Eurofighter Typhoon pictured during a flight demonstration at the Farnborough Air Show.

10

NEW ORDER, NEW CHALLENGES

One event changed, at a stroke, the fortunes of the RAF: the collapse of the Berlin Wall in November 1989. Within two years, both the Warsaw Pact and the Soviet Union had ceased to exist. The Western nations immediately planned to reduce or dismantle those military forces which had been associated with countering the Eastern bloc threat. These cutbacks might have been more wide-ranging had not Saddam Hussein decided to annex the Gulf state of Kuwait by military invasion on 2 August 1990. A coalition of nations (initially consisting of several Arab countries, the USA, Britain and France) determined first to defend the sovereignty of neighbouring Saudi Arabia, and then reconquer Kuwait. The defensive phase, Operation DESERT SHIELD, involved a build-up of 53,000 UK personnel, including a very substantial RAF contingent.

Saddam Hussein's miscalculation in the First Gulf War was that those assets deployed against him had been, as little as a year before, arrayed against the Warsaw Pact. Had the Cold War still been ongoing, Iraq might well have got away with taking Kuwait as her nineteenth province, because the West's military forces could not have been in two places at once and resisting Soviet aggression would have remained the priority. Similarly, had he waited a further year or two before invading Kuwait, the post-Cold War peace would have seen many of the personnel that deployed against him pensioned off, their units disbanded and their equipment sold or scrapped. Like many dictators, Saddam Hussein, lacked the geo-political understanding to recognize this.

Air Vice Marshal Andrew "Sandy" Wilson was appointed Air Commander British Forces Middle East, based in Riyadh, and deputy to the UK Joint Force Commander, Lieutenant-General Sir Peter de la Billière. The UK mission, known as Operation GRANBY, saw Tornados and Jaguars painted "desert pink" and immediately despatched to conduct defensive counter-air operations. Most of the RAF's strategic aircraft (the Hercules, VC-10 and Tristars) began moving personnel and supplies to the Gulf. In all, they would deploy 25,000 passengers and 31,000 tonnes of freight to the region. Meanwhile, Nimrods undertook intelligence-gathering and surveillance missions to maintain an economic embargo of Iraq and Kuwait, while RAF tankers supported the

whole effort, off-loading 13,000 tonnes of fuel, including the air-to-air refuelling of foreign coalition aircraft.

In effect, these roles – in type, intensity and endurance – were everything the RAF had trained to do for decades, albeit in West Germany against the Warsaw Pact. The well-oiled, adaptable machine had instead to cope with excessive heat – and the corrosive effect of sand. Almost 20 per cent of the RAF's deployment comprised the RAF Regiment, who undertook NBC (nuclear, biological, chemical) monitoring, and the air and ground defence of airbases, using its Rapier short-range air defence missile systems. Whilst the UN attempted to negotiate the departure of Iraqi forces from Kuwait, the coalition built up their strength. An ultimatum to leave Kuwait by 15 January 1991 was ignored, and Operation DESERT STORM – the invasion of Kuwait – began in the early hours of 17 January.

The RAF and Coalition air forces were now emerging into a modern world in which much of the weaponry was significantly more sophisticated than in previous conflicts. A large number of RAF aircraft were involved in this phase of DESERT STORM,

attacking a variety of Iraqi ground installations and command posts. Flying at night, RAF and Saudi Tornados dropped a number of new missiles and ordnance, including the JP233 (which contained runway cratering sub-munitions), whilst by day Jaguars attacked supply dumps, enemy missile sites, artillery positions and shipping. Buccaneers equipped with Pave Spike Laser Target Designators enabled Tornados to launch smart weapons against specific targets, bridges and hardened aircraft shelters (HAS). Later on in the campaign, Tornados equipped with Thermal Imaging Airborne Laser Designators (TIALD), were able to guide their wingmen's Paveway laser-guided bombs. For the first time in war, precision-guided munitions (PGMs), both bombs and missiles, played a decisive role, and critically weakened the world's fourth-largest army.

Opposite / Jaguar and Tornado fighter-bombers are liveried in sand-coloured camouflage to provide extra concealment for their desert operations during Operation GRANBY in 1991.

Below / Men of the RAF Regiment man a Rapier air-defence battery in Bahrain during Operation GRANBY, the British contribution to the first Gulf War in 1991.

After the intensive air campaign, which suppressed enemy air defences, neutralized any threat from the Iraqi air force and paralyzed their ground communications, the Coalition launched their land assault on 24 February 1991. RAF Chinooks and Pumas flew 900 sorties, mainly in support of the 1st (UK) Armoured Division. Iraqi ground forces had been so degraded from the air, they were soon in headlong retreat. The relentless air campaign continued to hammer their withdrawal up the Basra Road, resulting in scenes similar to the wrecked German forces trying to escape the Normandy pocket back in 1944. The land campaign ceased at 0500 hours on 28 February, after only 100 hours of combat and without the Coalition following the retreating Iraqis back across the border; at the time, there was a reluctance to take the campaign beyond the Kuwaiti borders.

The Gulf War had largely been decided in the air, even though it was sea-launched missiles that dominated the news coverage, despite their representing only a fraction of the Coalition fire-power. In all, the Coalition lost 75 aircraft; of the 158 RAF aircraft deployed on over 6,000 sorties, six Tornados were lost, with five aircrew killed and seven captured, a remarkable exchange rate compared with earlier wars and an illustration of the effectiveness of modern technologies. Later repatriated by the Red Cross, Flight Lieutenants John Peters and John Nichol of XV Squadron were memorably shown on Iraqi television, bloodied and beaten after their aircraft had been downed by a shoulder-launched SA-14.

After DESERT STORM had ceased, Coalition forces continued to enforce a no-fly zone over Iraq until 2003, and their operations included a British air initiative run out of Turkey to protect Kurdish civilians fleeing their homes in northern Iraq in the aftermath of the 1991 war, and the delivery of humanitarian aid. Throughout this period, RAF Jaguars and Tornados participated in daily patrols in the airspace over Iraq and conducted regular bombing campaigns targeting Iraqi anti-aircraft defences. In December 1998, frustrations over Iraqi non-compliance with UN Security Council resolutions and their hindrance of UN-authorized weapons inspectors flared up into another fully-fledged, four-day air campaign, Operation DESERT FOX. The RAF's contribution included four Jaguar aircraft based in Turkey patrolling Iraq's northern zone and 18 Tornados over the southern zone, flying out of Kuwait and Saudi Arabia.

Global Positioning System (GPS) units, then in their infancy, were used to great effect by the Coalition in the Gulf War and afterwards, while just as important was the contribution made by the Airborne Warning and Control System (AWACS) aircraft. The RAF had ordered seven Boeing E-3D Sentry AWACS in 1987, with deliveries starting in 1990. These purchases came about following the cancellation of the British Aerospace Nimrod project to replace the Avro Shackleton from the 1980s. Although the RAF operated 46 Nimrod MR1s from 1969 and 35 MR2s from 1979, technological challenges and cost overruns in developing the upgraded MRA4 led to the project being cancelled in 2010 and not replaced. Furthermore, in 2006, a Nimrod MR2 was lost while intelligence-gathering over Afghanistan, killing all 14 servicemen on board – the largest loss of UK military personnel in a single event since the Falklands War. This was not the result of ground fire, as the Taliban claimed, but was related to a fuel leak and possibly the age of the aircraft, which by then was nearly 40 years old. At any rate, it was a tragedy that influenced the decision to retire the Nimrod MR2s. A maritime capability gap opened, and by 2015 Eurofighter Typhoons from 2 Squadron were training with Type 45 destroyers in an Air-Maritime Integration (AMI) role – an admission that such expertise had been neglected. Not until 2016 was the reconnaissance gap to counter a renewed

Above / Two Nimrod R1 electronic intelligence aircraft of 51 Squadron based at RAF Waddington, Lincolnshire, 17 July 1995.

Opposite, top / A Sentinel R1 surveillance aircraft of 5 (AC) (Army Cooperation) Squadron. These aircraft can detect and recognize moving, static and fixed targets on the ground and are capable of operating for over nine hours at a time.

Opposite, below / Image of the starboard side of a 14 Squadron Shadow R1 Aircraft based at RAF Waddington.

Russian maritime threat addressed with the purchase of nine Boeing P-8 Poseidons. The first, however, was not delivered until November 2019.

Shrouded in secrecy until recently, the RAF did operate three Nimrod R1s, an electronic intelligence gathering (ELINT) variant. These were eventually pensioned off in 2011 and replaced by a similar number of Boeing RC-135W Rivet Joint aircraft – another off-the-shelf American purchase – and operated by 51 Squadron from Waddington. Their intelligence role is supported by eight Shadow R1s (a modified Beechcraft Super King Air) of 14 Squadron, and five Raytheon Sentinel R1s (based on the Bombardier Global Express ultra-long-range business jet) aircraft, with 5 Squadron. Both types have been such a force-multiplying component of operations in Afghanistan that their in-service life has been extended.

Meanwhile, the Gulf War had demonstrated that precision weapons such as laser-guided bombs had greatly enhanced the effectiveness of air power and showed that even targets like tanks and personnel carriers could now be accurately picked off and destroyed. Smart weapons had become capable of tactical interference in the modern battle space as well as usable against targets of fundamental importance, such as enemy headquarters, missile stations and air assets. This represented a massive leap forwards. At the same time, accidental – collateral – damage to civilians and their buildings had markedly reduced. As an unnamed Coalition pilot observed, "Frankly, I would not have believed ten years ago that we could hit the targets we are routinely destroying today. War has changed completely, for better or worse. The side with the smartest gadgets will always win."

Although the Americans in Vietnam and the Israelis in the 1973 (Yom Kippur) War had used unmanned reconnaissance aircraft with great success, it was the maturing and miniaturization of computer technologies in the 1980s that prompted the US to acquire and develop some Israeli-designed Unmanned Aerial Vehicles (UAVs), more commonly known as "drones", experimenting with concepts similar to those that governed smart weapons. Thus, the first "UAV

war" was the 1991 Gulf conflict, where at least one American UAV was airborne at all times during DESERT STORM.

After their service in the Gulf War had successfully demonstrated their utility, many defence forces worldwide studied the military applications of UAVs, but none more so than the United States. Closely monitoring the UAV activities of its closest ally, in October 2007 the RAF first acquired MQ-9A Reaper drones for operations in Afghanistan. In UK military parlance such devices are Remotely Piloted Aircraft Systems (RPAS), with the emphasis on the fact that an intelligent human pilot, rather than a computer, governs the UAV process. Initially, they were intended only to provide a 24-hour, all-weather Intelligence, Surveillance, Target Acquisition and Reconnaissance (ISTAR) capability. Like the Americans, the British soon realized that the platform could be enhanced to strike at chosen targets and provide close air support to ground troops; in May 2008, the RAF's Reaper fleet began missions armed with GBU-12 227kg (500lb) Laser Guided Bombs and up to four AGM-114 Hellfire missiles.

Unmanned Combat Air Vehicle (UCAV) or RPAS are evolving in importance in all of the world's air forces, and the RAF's Reaper capability has grown into two squadrons, 39 (at Creech Air Force Base, Nevada) and 13 Squadron, based at Waddington. Each Reaper is launched by a local ground crew before being handed over to a mission crew located at either Creech or Waddington, who pilot the drone via a secure satellite link; control is once again passed to the local crew for landing. Working closely with the intelligence communities, they can be deployed not just against conventional enemy forces, but to combat terrorists, pirates or drug and people smugglers.

The Coalition's success in the 1991 Gulf War reflected a change of US airpower doctrine, which the United Kingdom also supported. The prevailing view throughout the 1980s held that air power should always play a subordinate role to ground operations, but USAF Colonel John A. Warden suggested an alternative. Irrespective of land-force requirements, Warden devised a concept of five "rings" of target sets that airpower should subdue. In order of importance, they were enemy leadership locations; key production centres; infrastructure nodes such as power stations; targets to affect and paralyze the population; and finally, the degradation of military forces. He argued that attacking each of these would have a strategic effect on an opponent. It was Warden who was behind the DESERT STORM air war campaign and his views have prevailed in USAF and RAF circles ever since. In many ways, his ideas were merely an extension of the doctrine and interdiction tactics of the Second World War.

The collapse of the Soviet Union, also ushered in an era of interventionism in the minds of Western politicians. In 1980 Marshal Tito, last of the great Second World War leaders, had died. He had led Yugoslavia since 1944, not as a single sovereign nation, but a federation of six republics, with uncertain borders drawn along ethnic and historical lines. Whereas Tito had used violence and repression to forcibly hold together a fractured, rather than a united, country, Slobodan Milošević (the president of Serbia) did the opposite. He exploited the fractures in the hope of making his republic dominant. He wanted nothing less than a Greater Serbia in place of the old Yugoslavia.

This resulted in widespread ethnic tensions and massacres, often against the Bosnian Muslim population, of an intensity and viciousness not seen since 1939–45. From 1992, the United Kingdom, together with other nations found itself increasingly drawn into UN-sponsored peacekeeping efforts in the region. Echoing aspects of the Berlin Airlift, C-130 Hercules from the Lyneham Wing airlifted humanitarian supplies into the besieged Bosnian capital of Sarajevo. Their nearly 2,000 sorties equated to over 25,000 tons of food and medical aid. The period also saw RAF Tornado F3s based in Italy and AWACS contributing to Operation

DENY FLIGHT, NATO's operation to enforce a no-fly zone over Bosnia, while some 2,500 UK personnel operated on the ground as part of UNPROFOR (the UN PROtection FORce).

As tension mounted in the region, August-September 1995 saw NATO bombing Serbian targets in Bosnia and Croatia: RAF Harriers equipped with laser-guided bombs (LGBs) flew 144 sorties, supported by Jaguars flying from Italy to guide the munition, in the first use of offensive air power in Europe since 1945. Tornados, Tristar tankers and AWACS also participated, amounting to 10 percent of the total Coalition sorties. The United Nations mission in Yugoslavia altered in December 1995 with the signing of the Dayton Peace Accord, which allowed a substantial NATO ground force of some 60,000 personnel to enter Bosnia and stabilize the region. This saw the white vehicles and blue helmets of UN peacekeepers replaced by the camouflaged headgear and armour of NATO's peace enforcers. Known for its first year as IFOR (the peace Implementation FORce), together with its air aspect it evolved in 1997 into SFOR, the Bosnian Stabilisation

Below / An MQ-9A Reaper Drone of 39 Squadron at Kandahar airbase in Afghanistan in 2012. Although launched in-theatre, the Reapers are controlled from Creech Air Force Base in the USA.

Above / A pilot of 6 Squadron climbs into his Jaguar GR3 in February 1994 during Operation DENY FLIGHT to enforce a no-fly zone against Serb forces in Bosnia.

Opposite, above / An RAF Chinook attached to SFOR, the NATO-led peacekeeping force assigned to Bosnia-Herzegovina between 1996 and 2004, takes off from Split.

Opposite, below / RAF Harrier GR7s take off from Gioia del Colle airbase in southern Italy in April 1999 as part of Operation ALLIED FORCE, a series of strikes against Yugoslav forces to protect Albanian Muslims in Kosovo.

FORce. By Dayton, the Serbs had already ceased fighting, though six Harriers based in Italy kept watch, supported by Jaguars, Tristars and a Sentry.

Milošević continued to menace the region, encouraging Serbs to settle in Kosovo and eject local ethnic Albanians. The civilian population there responded by forming the Kosovo Liberation Army (KLA), who were effectively fighting an undeclared war with Serbian forces by the summer of 1998. With NATO overflights, including four RAF Jaguars, failing to subdue the Serbs, in March 1999, Operation ALLIED FORCE was launched. This evolved into a 78-day bombing campaign to halt the ethnic cleansing of Albanian Muslims, with a contribution of some 50 aircraft by the RAF. It was a significant campaign for the RAF, where 25 per cent of munitions dropped were precision-guided.

British forces gained more experience in interventionist missions overseas in 2000. The West African state of Sierra Leone had been consumed by civil war begun since 1991. British troops were first deployed in May 2000 to evacuate their own and other foreign nationals. In the aftermath, a Short Term Training Team (STTT) of United Kingdom forces remained to train and rebuild the Sierra Leone Army, in order to recreate security in the troubled state. However a militia group, the West Side Boys, refused to integrate into the reconstituted Sierra Leone Army and began operating as bandits from abandoned villages. A law unto themselves and freely taking drugs, they eventually captured a patrol of UK soldiers and held them hostage. RAF Chinooks and other aircraft helped deploy SAS and other troops to free the hostages and destroyed the gang as a threat in Sierra Leone. This small operation was highly successful and sent a message further afield that the Labour government under Tony Blair was prepared to use considerable force to back up humanitarian intervention, something the initial years in Bosnia had lacked.

Opposite / An RAF Chinook lands on the deck of HMS *Illustrious* in June 2000 during the Operation PALLISER mission to assist UN peacekeepers in Sierra Leone and train the Sierra Leonean army.

Below / Marines of 42 Commando disembark from a Chinook helicopter at Lungi airport, just north of the Sierra Leonean capital, Freetown, during Operation PALLISER.

eanwhile, it was obvious that Saddam Hussein remained as a malign force in the Middle East, with the United States determined to deal with him. Enlisting several Coalition partners including the United Kingdom, Operation IRAQI FREEDOM (Operation TELIC for United Kingdom armed forces) was unleashed on 20 March 2003. There was an aerial bombardment phase, but this time it coincided with the ground offensive, which was spearheaded by helicoptered Special Forces penetrations. Many media pundits, and no doubt the Iraqis themselves, expected an extended "shock and awe" air campaign, as in 1991: as ever, surprise is a vital principle of waging war.

Militarily, the Iraqi armed forces posed a viable threat, but the main hazard to the Coalition air forces was the Iraqi air-defence infrastructure, which was extensively targeted. The invasion of Iraq was one of the largest and most ambitious British missions since the Second World War. This was reflected by the RAF's participation. Some 206 aircraft (as it happened, with an almost equal proportion of fixed-wing to rotary-wing) and 8,000 air-force personnel had been deployed by the start of the invasion, amounting to some 30 per cent of the RAF's available strength. Moreover, the UK National Contingent Commander, co-located with the United States, was an airman, Air Marshal Sir Brian Burridge. He had overseen the targeting process in Kosovo and was very aware of London's desire

Above / Air Marshal Brian Burridge, overall commander of British forces during Operation TELIC, the 2003 invasion of Iraq, answers questions during a press conference in Doha, Qatar.

Below / An RAF Tornado GR4 undergoes maintenance at a Middle East base during Operation TELIC. Tornados remained committed to Iraq until 2009.

Opposite / Hercules transport plane arrives back at RAF Akrotiri in August 2014 after dropping supplies to towns in northern Iraq besieged by Islamic State fighters.

to keep Iraqi civilian casualties to a minimum. His view was shared by RAF component commander, Air Vice Marshal Glenn Torpy.

The invasion phase was achieved in a matter of weeks but there was confusion among the Coalition as to how to administer the defeated nation afterwards. As a result, the liberators came to be seen as occupiers and parts of Iraq drifted into insurgency. After a brief window of exuberance, Coalition forces were obliged to go over to the defensive and to begin to fight a counter-insurgency campaign. The initial Iraq operations also witnessed several RAF innovations, one of which was the use of Storm Shadow all-weather stand-off air-to-ground missiles, which could be launched at targets 240 kilometres (150 miles) distant.

RAF structures had also been altered, which benefited the Iraq mission and subsequent operations. A Joint Helicopter Command was created in October 1999, to operate and maintain the battlefield helicopters of the navy, army and air force; for Iraq, the Joint Helicopter Force (JHF) Iraq operated RAF Chinooks, Pumas and Merlins. The Joint Force Harrier (JFH), established in 2000 as part of Strike Command, did the same for the Royal Navy's two Sea Harrier squadrons and the RAF's four Harrier GR7 squadrons, overseeing their activities from carriers, RAF stations and deployed air bases. Later known as the Joint Strike Wing, its first and largest deployment, like that of the JHF, was in Iraq, followed by Afghanistan.

Apart from sustainment of the operation, the RAF's chief contribution later on in Iraq was in intelligence gathering and reconnaissance. RAF losses were few, but in January 2005, a C-130 was shot down north of Baghdad with the loss of all on board – eight crew from No. 47 Squadron based at Lyneham, another RAF

serviceman and one soldier. The following year, Flight Lieutenant Sarah-Jayne Mulvihill became the first RAF servicewoman to be killed on operations when her Lynx helicopter was shot down over Basra; four colleagues, including Wing Commander John Coxen, also perished in the same attack.

Sustainment of TELIC was predominantly carried out by C-17 and C-130 aircraft. The four-engined, turbo-prop Lockheed C-130, with its distinctive tail ramp, first emerged from a design requirement following the Korean War and entered US service in 1956. It was used extensively in Vietnam and all subsequent campaigns, and first flew in RAF colours in 1966. Since then, more than 90 of various marks have passed through RAF hands, and its current strength of 24 C-130Js are all based at Brize Norton. Most military personnel deploying anywhere will have travelled by C-130 in their service careers. They are being slowly replaced by the Airbus A400M Atlas, of which 22 are on order, along with nine Airbus A330 Voyagers, which form the new tanker fleet. The multi-role Voyagers are capable of aeromedical evacuation, as well as transporting personnel and freight, besides air-to-air refuelling.

Much of the Hercules fleet had been based at Lyneham, which became the gateway for personnel deploying to and from the Balkans, Iraq and Afghanistan. The station was also where the bodies of those killed on operations were received and then conveyed through the nearby town of Wootton Bassett, where crowds routinely lined the streets in tribute to the fallen. When Lyneham closed and its aircraft moved to Brize Norton, in recognition of the town's dignified support to the military community, it was renamed Royal Wootton Bassett by Queen Elizabeth II in 2011.

The long history of RAF humanitarian aid and disaster relief missions has also rested in recent decades on the C-17 and C-130 fleets. In 1984–85, the RAF airlifted 32,000 tons of grain and medical supplies to help alleviate famine in Ethiopia. Other mercy missions were flown to Chile (1991), Turkey and the Caribbean (both 1992), Somalia (1993), Rwanda (1994), Angola and Montserrat (both 1995) and Mozambique (2002), whilst five Chinooks were deployed by C-17 from the UK to assist in the wake of the 2005 earthquake in Northern Pakistan in Operation MATURIN. RAF support was also vital to the UK-Irish-Canadian response to the Ebola outbreak in West Africa in March 2014.

Opposite / Two Hercules C1 transport aircraft in flight in 1970. Introduced in the 1967, the Hercules remained the backbone of RAF transport operations until its retirement in 2011.

Above, top / Royal Air Force movements staff unload essential kit from a 99 Squadron C-17 at Cebu airport on 16 November 2013.

Above / Crewmembers relax in a C-17 on their way to deliver aid to the Philippines following the devastation caused by Typhoon Haiyan in November 2013.

Following the world-changing events of 11 September 2001, with the hijacking of domestic airliners in the US and the attacks on the Twin Towers of the World Trade Center by Al-Qaeda, the United States was determined to pursue the mastermind of the attacks, Osama bin Laden. He was based with the sympathetic Taliban, a Sunni extremist sect then governing Afghanistan. Air attacks began against Al-Qaeda almost immediately under Operation ENDURING FREEDOM. A NATO-led security mission in Afghanistan was established by the UN Security Council in December 2001, called the International Security Assistance Force (ISAF).

ISAF was initially charged with securing Kabul and the surrounding areas from the Taliban, Al-Qaeda and factional warlords, but in 2003, its mission was expanded throughout Afghanistan. It became increasingly involved in ever more intensive combat operations in the southern and eastern provinces, a role that was reflected by the amount of airpower involved in sustaining the mission, evacuating casualties, and providing close air support and reconnaissance. Coalition partners included the United States, United Kingdom and other NATO member states, some of whom contributed military aircraft. ISAF ceased combat operations and was disbanded in December 2014, although some personnel remained behind in an advisory role.

RAF Canberras, Nimrods and E-3 Sentrys were immediately involved in the Coalition effort against the terrorists. Meanwhile, C-17s and C-130s started moving personnel, supplies and equipment into theatre, supported by the RAF's tanking fleet. As the intensity and tempo of operations – and associated casualties – grew, there was pressure to expand the United Kingdom commitment. General David Richards, the future Chief of the

Defence Staff, observed that it "was the heaviest persistent combat the British Armed Forces had experienced since the Korean or the Second World War". By 2007, the UK commitment had expanded to over 7,000 personnel, and more aircraft. Throughout these operations, RAF Harriers, Army Air Corps Apache AH1s and RPAS provided aerial reconnaissance and fire support.

As the campaign gathered pace, with an ever-increasing number of Coalition troops on the ground, RAF support helicopters – Chinooks and Merlins – became increasingly key in the role of casualty evacuation (Casevac), due to their size and carrying capacity. The tandem-rotored Chinooks, normally based at Odiham, have worn RAF colours since entering service in 1980, serving in every campaign since then. One, *Bravo November*, was the sole survivor of five deployed to the Falklands in 1982. *Bravo November* afterwards had a remarkable service life, operating in Iraq, Northern Ireland and Afghanistan. The Chinook fleet has gradually climbed in size – and

importance – with a planned size of 60 aircraft.

The Merlins first deployed in 2003 to the Balkans in support of UN Forces, but in Afghanistan they and the Chinooks began carrying Medical Emergency Response Teams (MERTs). Crewed by serving medical staff and their equipment, the helicopters effectively became flying emergency wards, where the wounded – many often horrifically so – were given a chance to live because the Merlins offered increasingly superb medical stabilizing procedures while airborne until more extensive aid could be reached. In the words of Flight Lieutenant Andy Smith, an instructor on the MERT course in 2010, "On Op HERRICK, we treat casualties no

Above / An RAF Chinook resupplies a patrol base against the backdrop of a sandstorm during Operation GLACIER 4 to attack a Taliban-held fort in Helmand Province, Afghanistan, in 2008.

Opposite / A Chinook lands Royal Marines of 42 Commando during Operation SILICA to relieve Royal Fusiliers holding the town of Nowi Zad in Helmand Province in 2008.

matter who they are – insurgents or coalition, children or women."

Joining these aircraft in Afghanistan were detachments of Harriers, flying from Kandahar, which was also home to the RAF's No. 904 Expeditionary Air Wing, C-130s and USAF aircraft. The largest British base in Afghanistan was Camp Bastion near Lashkar Gah in Helmand Province. Gradually expanded from 2007 to

reach a garrison of 32,000, with its own control tower and runway capable of receiving C-17s, alongside a fully-equipped military hospital, it housed No. 903 Expeditionary Air Wing. Chinooks and Merlins were also based there, with security provided by the RAF Regiment. It was handed over to the Afghan Security Forces in 2014.

Meanwhile, beginning in March 2011, the UK had also undertaken a military intervention in Libya. Largely air-force-led, this was an international effort to protect the Libyan population, and authorized by the UN Security Council. The Libyans had risen against their ruler Colonel Muammar Gaddafi as part of the wider Arab Spring which affected most Middle Eastern states. Tornados – flying from Marham and refuelled three times en route – released Storm Shadow weapons at military targets, supported by Sentry, Sentinel and VC-10s out of Cyprus. As the mission grew, two RAF Expeditionary Air Wings were formed, 906 in Italy and 907 at Akrotiri.

Conceived as far back as the mid-1980s, and beset with many technological and financial challenges, the twin-engined, multi-role Eurofighter Typhoon had been originally ordered as an air defence fighter by the United Kingdom, Germany, Italy and Spain. The Typhoon remains a fine aircraft, but demonstrates just how complex modern war jets have become and how difficult they are to develop and then produce; RAF aircraft have come a very long

way in 100 years of operations. It finally entered RAF service with 3 Squadron at Coningsby in 2005. Initially deployed to the Falklands, and engaged in QRA missions against Russian aircraft, RAF Typhoon FGR4s had their combat debut over Libya, with ten aircraft flying alongside 16 RAF Tornados and four Typhoons of the Italian Air Force. They performed aerial reconnaissance and ground-attack missions using Paveway LGBs and Brimstone air-to-ground missiles.

The NATO Coalition effort also included substantial contributions from France and Canada, eventually expanding into 19 (including some non-NATO) nations, but noticeably the USA did not lead. They shared command, and provided vital intelligence, but this was the first time the world's largest power did not dominate an overseas military intervention. Some 26,000 coalition sorties were flown, which destroyed or damaged an estimated 6,000 regime targets. With the death of Gaddafi on 20 October 2011, the mission was wound down, although Libya has still not fully emerged from its post-Gaddafi chaos.

The Arab Spring that unsettled Libya also saw an insurrection against Syria's ruler, Bashar al-Assad. Taking advantage of the resultant uncertain internal security of both Syria and neighbouring Iraq, a radical, fundamentalist Sunni terrorist group, the Islamic State of Iraq and the Levant (ISIL) took control of territories in the border regions of both states. Also known as IS (Islamic State), ISIS (Islamic State in Iraq and Syria) or Daesh (a derogatory Arab acronym), the group brutally enslaved or killed anyone it disliked, including the local Yazidi Kurdish community in the north-western Iraqi town of Sinjar.

In August 2014, as part of an international coalition force, RAF C-130s began dropping humanitarian aid to fleeing Yazidis and airlifting those refugees in danger. These missions were extended in October 2014 to include surveillance flights over Syria, using Tornados and the new RC-135W Rivet Joints on their first operational deployment, flying out of Al Udeid airbase in Qatar. Reconnaissance was supplemented by Sentry, Sentinel, Shadow and Reaper fleets. Airstrikes against ISIL in Iraq were approved by the British Parliament in September following the execution of a British hostage. Operation SHADER, as it was called, was further extended in December 2015, in the wake of ISIS terrorist attacks in France, to include airstrikes in Syria – something the Prime Minister, David Cameron, had tried to achieve in 2013, but was prevented from doing so by a hostile House of Commons. Active operations over Syria began within hours, with Tornados and Typhoons attacking ISIS targets. The intensity of air operations increased and, by 2016, the RAF's commitment to SHADER had outstripped its involvement in Iraq and Afghanistan, with nearly 3,000 sorties, including over 2,000 strike missions.

Opposite / An RAF Puma disembarks troops during a 2009 exercise in Kenya as preparation for deployment to Afghanistan.

Below / Aircraft of 903 Expeditionary Air Wing gathered at Akrotiri in Cyprus in 2015 as part of Operation SHADER aimed against Islamic State in Iraq (and, later, in Syria).

An RAF Tornado GR4 returns to RAF Akrotiri after a raid on
Islamic State positions during Operation SHADER.

On the eve of its centenary, the RAF witnessed the emergence of many new threats, some from insurgent terrorist groups like ISIS. For example, in 2013, two C-17s and a Sentinel supported French operations against radical Muslim militias in Mali. Similar radical Sunni sects also threaten regional destabilization, like Boko Haram in the West African states of Nigeria, Mali, Chad and Niger, whilst Al Shabaab threatens similar disruption in Somalia, Somaliland, Eritrea and Kenya. Elsewhere, a militarily resurgent Russia – active on many fronts, from the Baltic to the Crimea and Syria – is a reminder that the RAF's traditional equipment and capabilities, with the professionalism and agility of its personnel, must not be eroded.

Multinational military operations will also be conducted in the domains of media, cyber and space, and the creation of the RAF-led Joint Forces Cyber Group in 2013, staffed by regulars and reservists, is indicative of the direction of future warfare. In 2014, Air Marshal Phil Osborn confirmed that the unit was dedicated to offensive and defensive cyber and electromagnetic warfare, and was benefitting from an increasingly generous financial investment.

New technologies are always knocking at the door, and besides cyber capability, the opportunities offered by stealth currently occupy much RAF attention. Some of it arrived in the RAF in June 2018 in the form of the Lockheed Martin F-35 Lightning.

This single-seat, all-weather stealth multi-role fighter had already entered USAF and USMC service, and the F-35B variant entered RAF service, based at Marham, just in time for the centenary, and 617 Squadron, perhaps the RAF's most famous, was the first to use this exciting new fighter jet. Others will equip the Royal Navy's two Queen Elizabeth-class carriers. It will be the UK's main attack aircraft until 2050, replacing the already-retired Harrier GR9s, and the Tornado GR4 fleet, which was retired in 2019.

With such an array of current and future threats, some fear that the distinctions between navy, army and air force are becoming increasingly blurred, as joint organizations take the lead. Nevertheless, the RAF has, perhaps more readily than its service comrades, signed up to the Whole Force Concept, which forges regulars and reservists, civil servants and contractors into a balanced organization. This provides added strength to tackle the challenges posed by the twenty-first century. With its strong ethos of respect, integrity, service and excellence, the Royal Air Force is well set to be propelled through its next century of service to the nation.

Opposite / Aircract perform a fly-past as the Royal Navy's new supercarrier HMS *Queen Elizabeth* sails into Portsmouth Harbour in August 2017.

Above / The RAF began to take delivery in 2012 of the Lockheed Martin F-35 Lightning, a new advanced fighter with stealth capabilities intended to replace its Harriers and Tornados.

INDEX

PICTURE CREDITS

The vast majority of photographs reproduced in this book have been taken from the collections of the Photograph Archive at the RAF Air Historical Branch. UK Crown Copyright.

Photographs from sources outside the Air Historical Branch, with the kind permission of the following:

8. The Print Collector/Print Collector/Getty Images, 9. Library of Congress, 10. Bettmann/Getty Images, 11. (top) Bob Thomas/Popperfoto/Getty Images, 12. (bottom) Marc Tielemans/Alamy Stock Photo, 14. Art Media/Print Collector/Getty Images, 16. (bottom) adoc-photos/Corbis via Getty Images, 18. Topical Press Agency/Hulton Archive/Getty Images, 19. SSPL/Getty Images, 24-25. akg-images/Osprey Publishing/London 1917–18 - The Bomber Blitz/Ian Castle, 26. (bottom) Public Domain, 35. Fremantle/Alamy Stock Photo, 39. (bottom) Fox Photos/Getty Images, 45. Antiqua Print Gallery/Alamy Stock Photo, 53. Derek Berwin/Fox Photos/Getty Images, 54. (left) Popperfoto/Getty Images, (right) Pictorial Press Ltd/Alamy Stock Photo, 65. Pictorial Press Ltd/Alamy Stock Photo, 73. Pictorial Press Ltd/Alamy Stock Photo, 101. (top) A. J. O'Brien/Fox Photos/Getty Images, (bottom) Fox Photos/Getty Images, 103. GraphicaArtis/Getty Images, 105. (top) Fox Photos /Hulton Archive /Getty Images, 106. (left) War Archive/Alamy Stock Photo, 113. B J Daventry/IWM via Getty Images, 127. Charles E. Brown/Royal Air Force Museum/Getty Images, 133. Bettmann/Getty Images, 137. Imperial War Museums, London (FLM 2340), 140. Trinity Mirror/Mirrorpix/Alamy Stock Photo, 142. (bottom) author collection, 150. (top) Science History Images/Alamy Stock Photo, 157. (top) SSPL/Getty Images, 162. Bentley Archive/Popperfoto/Getty Images, 163. Hulton Archive/Getty Images, 177. Bettmann/Getty Images, 178. ZUMA Press, Inc./Alamy Stock Photo, 182. (centre) Popperfoto/Getty Images, 184. Keystone Pictures USA/Alamy Stock Photo, 197. Rob McEwan/Moment/Getty Images, 208. (top) Carlo Allegri/Getty Images, 218. Carl Court/Getty Images

Every effort has been made to acknowledge correctly and contact the source and/or copyright holder of each picture, and Welbeck Non-fiction Ltd apologizes for any unintentional errors or omissions, which will be corrected in future editions of this book.